# JUMP Math 2.2

**Book 2** Part 2 of 2

## Contents

jump math™

MULTIPLYING POTENTIAL.

**JUMP Math**
One Yonge Street, Suite 1014
Toronto, Ontario M5E 1E5
Canada
www.jumpmath.org

Writers: Dr. Heather Betel, Julie Lorinc, Dr. John Mighton
Consultants: Dr. Anna Klebanov, Dr. Sindi Sabourin
Editors: Megan Burns, Liane Tsui, Natalie Francis, Julia Cochrane, Jackie Dulson, Neomi Majmudar, Una Malcolm, Rachelle Redford, Rita Vanden Heuvel
Layout and Illustrations: Linh Lam, Fely Guinasao-Fernandes, Sawyer Paul, Marijke Friesen, Pam Lostracco
Cover Design: Blakeley Words+Pictures
Cover Photograph: © iStockphoto.com/Michael Valdez

ISBN 978-1-927457-38-2

Second printing July 2017

Printed and bound in Canada

# Welcome to JUMP Math

Entering the world of JUMP Math means believing that every child has the capacity to be fully numerate and to love math. Founder and mathematician John Mighton has used this premise to develop his innovative teaching method. The resulting resources isolate and describe concepts so clearly and incrementally that everyone can understand them.

JUMP Math is comprised of teacher's guides (which are the heart of our program), interactive whiteboard lessons, student assessment & practice books, evaluation materials, outreach programs, and teacher training. The Common Core Editions of our resources have been carefully designed to cover the Common Core State Standards. All of this is presented on the JUMP Math website: **www.jumpmath.org**.

Teacher's guides are available on the website for free use. Read the introduction to the teacher's guides before you begin using these resources. This will ensure that you understand both the philosophy and the methodology of JUMP Math. The assessment & practice books are designed for use by students, with adult guidance. Each student will have unique needs and it is important to provide the student with the appropriate support and encouragement as he or she works through the material.

Allow students to discover the concepts by themselves as much as possible. Mathematical discoveries can be made in small, incremental steps. The discovery of a new step is like untangling the parts of a puzzle. It is exciting and rewarding.

Students will need to answer the questions marked with a ⬜ in a notebook. Grid paper notebooks should always be on hand for answering extra questions or when additional room for calculation is needed.

# Contents

## Unit 4: Operations and Algebraic Thinking: Unknowns in Subtraction

## Unit 5: Number and Operations in Base Ten: Addition Using Place Value

## Unit 6: Number and Operations in Base Ten: Subtraction Using Place Value

# Unit 7: Measurement and Data: Measuring Length in Metric Units

# Unit 8: Measurement and Data: Measuring and Operations

# PART 2

# Unit 1: Operations and Algebraic Thinking: Compare Problems

# Unit 2: Number and Operations in Base Ten: Three-Digit Numbers

## Unit 3: Operations and Algebraic Thinking: Two-Step Word Problems

## Unit 4: Number and Operations in Base Ten: Strategies for Large Numbers

## Unit 5: Measurement and Data: Measuring in US Customary Units

# Unit 6: Measurement and Data: Time

# Unit 7: Measurement and Data: Money

# Unit 8: Geometry: Shapes

# Unit 9: Measurement and Data: Graphs

# OA2-44 Subtracting by Using 10

Show the subtraction.
○ Take away ⬜ first. Then take away as many ☐ as needed.

**1.**

$$16 - 8$$

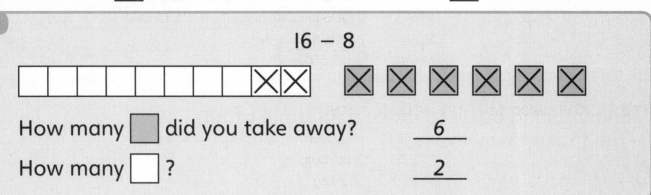

How many ⬜ did you take away?    6

How many ☐ ?    2

**2.**

$$15 - 6$$

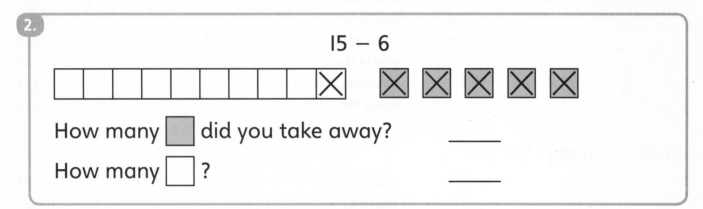

How many ⬜ did you take away?    ____

How many ☐ ?    ____

**3.**

$$14 - 7$$

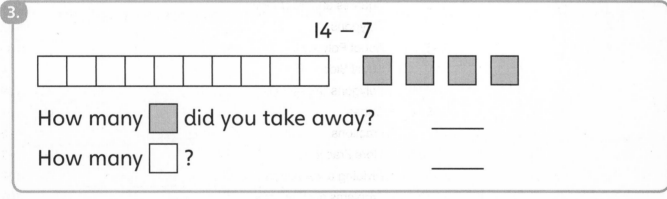

How many ⬜ did you take away?    ____

How many ☐ ?    ____

**4.**

$$17 - 9$$

How many ⬜ did you take away?    ____

How many ☐ ?    ____

$$7 = 1 + 6 \qquad 8 = 1 + 7 \qquad 9 = 1 + 8$$
$$7 = 2 + 5 \qquad 8 = 2 + 6 \qquad 9 = 2 + 7$$
$$7 = 3 + 4 \qquad 8 = 3 + 5 \qquad 9 = 3 + 6$$
$$\phantom{7 = 3 + 4} \qquad 8 = 4 + 4 \qquad 9 = 4 + 5$$

☐ Show taking away to make a multiple of 10.
   Then take away the rest.

**5.**    12    −    8

= 12 − __2__ − __6__

**6.**    11    −    6

= 11 − __1__ − ____

**7.**    13    −    7

= 13 − ____ − ____

**8.**    22    −    5

= 22 − ____ − ____

**9.**    45    −    8

= 45 − ____ − ____

**10.**    77    −    8

= 77 − ____ − ____

☐ Take away to make a multiple of 10. Then take away the rest.

**11.**    12    −    8

= 12 − __2__ − __6__

= __10__ − __6__

= __4__

**12.**    15    −    6

= 15 − __5__ − ____

= __10__ − ____

= ____

**13.**    14    −    7

= 14 − ____ − ____

= ____ − ____

= ____

**14.**    37 − 9

**15.**    53 − 6

**16.**    92 − 9

**17.**    12 − 4

**18.**    64 − 6

**19.**    86 − 8

# OA2-45 More and Fewer

☐ Draw ◯ or △ to show which is fewer.

| 1. | Which is fewer? |
|---|---|
| ◯◯◯◯◯◯ △△△△ | △ |
| ◯◯ △△△△△ | |
| △△△△△△△ ◯◯◯ | |
| △△△△ ◯◯◯◯◯◯◯ | |

☐ Circle **more** or **fewer** △.

| 2. | More or fewer △? |
|---|---|
| ◯◯◯ △△△△△△ | (more) / fewer |
| ◯◯◯◯ △△ | more / fewer |
| △△△△△ ◯◯◯◯◯◯◯◯ | more / fewer |
| △△△△△△△△ ◯◯◯◯◯◯ | more / fewer |

☐ Draw ◯ or △ to show which is fewer.
☐ Write how many fewer.

| 3. | | Which is fewer? | How many fewer? |
|---|---|---|---|
| ◯ ◯ ◯<br>△ △ △ △ △ | | ◯ | 2 |
| ◯ ◯ ◯ ◯ ◯ ◯<br>△ △ | | | |
| △ △ △ △<br>◯ ◯ ◯ ◯ ◯ ◯ ◯ | | | |
| △ △ △ △ △ △ △ △ △<br>◯ ◯ ◯ ◯ | | | |

☐ Draw ◯ or △ to show which is more.
☐ Find how many more.

| 4. | | Which is more? | How many more? |
|---|---|---|---|
| 5 ◯ or 8 △ | | △ | _8_ – _5_ = _3_ |
| 12 ◯ or 7 △ | | | ___ – ___ = ___ |
| 14 ◯ or 12 △ | | | ___ – ___ = ___ |
| 9 ◯ or 12 △ | | | ___ – ___ = ___ |
| 11 ◯ or 6 △ | | | ___ – ___ = ___ |
| 10 ◯ or 15 △ | | | ___ – ___ = ___ |

Operations and Algebraic Thinking 2-45

☐ Underline who has fewer.
☐ Find how many fewer.

| 5. | How many fewer? |
|---|---|
| <u>Tess</u> has 5 shells.<br>Jack has 8 shells. | $\underline{\ 8\ } - \underline{\ 5\ } = \underline{\ 3\ }$ |
| Emma has 17 shells.<br>Ted has 7 shells. | ___ − ___ = ___ |
| Ray has 14 shells.<br>Grace has 18 shells. | ___ − ___ = ___ |
| Ava has 9 shells.<br>Tony has 15 shells. | ___ − ___ = ___ |
| Nina has 11 shells.<br>Ivan has 4 shells. | ___ − ___ = ___ |

☐ Find how many more or fewer.

**6.**

Yu has 8 raisins. Bill has 12 raisins.

How many more raisins does Bill have? ___ − ___ = ___

**7.**

Alex has 13 raisins. Ethan has 2 raisins.

How many fewer raisins does Ethan have? ___ − ___ = ___

**8.**

Clara has 13 raisins. Abdul has 2 raisins.

How many more raisins does Clara have? ___ − ___ = ___

# OA2-46 Compare Using Pictures

☐ Circle who has more. Underline who has fewer.

☐ Draw triangles for Rosa.

**1.**

(Rosa) has 3 more △ than <u>Sam</u>.

| Sam | △ | △ | △ | △ | △ | | | |
|------|---|---|---|---|---|---|---|---|
| Rosa | △ | △ | △ | △ | △ | △ | △ | △ |

**2.**

Rosa has 3 fewer △ than Sam.

| Sam | △ | △ | △ | △ | △ | | | |
|------|---|---|---|---|---|---|---|---|
| Rosa | | | | | | | | |

**3.**

Rosa has 1 more △ than Sam.

| Sam | △ | △ | △ | △ | △ | | | |
|------|---|---|---|---|---|---|---|---|
| Rosa | | | | | | | | |

**4.**

Rosa has 2 more △ than Sam.

| Sam | △ | △ | △ | △ | △ | | | |
|------|---|---|---|---|---|---|---|---|
| Rosa | | | | | | | | |

**5.**

Rosa has 4 fewer △ than Sam.

| Sam | △ | △ | △ | △ | △ | | | |
|------|---|---|---|---|---|---|---|---|
| Rosa | | | | | | | | |

◯ Draw triangles for Beth.
◯ Circle how many triangles Beth has.

**6.**

| Greg has 5 △. | △ | △ | △ | △ | △ | | | | | | $\boxed{5 + 2}$ |
| Beth has 2 more △ than Greg. | △ | △ | △ | △ | △ | △ | | | | | 5 − 2 |

**7.**

| Greg has 3 △. | △ | △ | △ | | | | | | | 3 + 2 |
| Beth has 2 fewer △ than Greg. | | | | | | | | | | 3 − 2 |

**8.**

| Greg has 6 △. | △ | △ | △ | △ | △ | △ | | | | 6 + 3 |
| Beth has 3 more △ than Greg. | | | | | | | | | | 6 − 3 |

**9.**

| Greg has 4 △. | △ | △ | △ | △ | | | | | | 4 + 3 |
| Beth has 3 fewer △ than Greg. | | | | | | | | | | 4 − 3 |

**10.**

| Greg has 7 △. | △ | △ | △ | △ | △ | △ | △ | | | 7 + 1 |
| Beth has 1 fewer △ than Greg. | | | | | | | | | | 7 − 1 |

COPYRIGHT © 2014 JUMP MATH: NOT TO BE COPIED. CC EDITION

# OA2-47 Comparing and Word Problems (1)

☐ Circle which is more. Underline which is fewer.
☐ Find how many circles.

**1.**

| △ | | How many ◯? |
|---|---|---|
| 5 | There are 3 more ◯ than △. | 5 + 3 = 8 |
| 6 | There are 3 fewer ◯ than △. | 6 – ___ = ___ |
| 4 | There are 2 more ◯ than △. | ___ = ___ |
| 9 | There are 6 fewer ◯ than △. | ___ = ___ |
| 15 | There are 4 fewer ◯ than △. | ___ = ___ |
| 13 | There are 5 fewer ◯ than △. | ___ = ___ |
| 9 | There are 4 more ◯ than △. | ___ = ___ |
| 21 | There are 5 more ◯ than △. | ___ = ___ |
| 27 | There are 3 fewer ◯ than △. | ___ = ___ |
| 39 | There are 6 fewer ◯ than △. | ___ = ___ |

○ Circle which is more. Underline which is fewer.

○ Add or subtract.

**2.**

Mona has 25 apples.

She has 12 more (pears) than apples.

How many pears does Mona have?

|   |   | 2 | 5 |
|---|---|---|---|
| + |   | 1 | 2 |
|   |   |   |   |

**3.**

Mark has 46 red flowers.

He has 17 fewer yellow flowers than (red) flowers.

How many yellow flowers does Mark have?

**4.**

Tim has 93 blocks.

He has 46 fewer balls than blocks.

How many balls does Tim have?

**5.**

Lynn drew 28 circles.

She drew 17 more triangles than circles.

How many triangles did Lynn draw?

☐ Underline **more** or **fewer**.

**1.**
Maria has 3 more stickers than Bob.
Bob has more / <u>fewer</u> stickers than Maria.

Maria has 3 fewer stickers than Bob.
Bob has more / fewer stickers than Maria.

**2.**
Carlos has 2 more stickers than Jen.
Jen has more / fewer stickers than Carlos.

Carlos has 2 fewer stickers than Jen.
Jen has more / fewer stickers than Carlos.

☐ Circle who has more. Underline who has fewer.
☐ Draw triangles for Rani.

**3.**

| (Kyle) has 2 more △ than <u>Rani</u>. | | | | | | | |
|---|---|---|---|---|---|---|---|
| Kyle | △ | △ | △ | △ | △ | | | |
| Rani | △ | △ | △ | | | | | |

**4.**

| Kyle has 2 fewer △ than Rani. | | | | | | | |
|---|---|---|---|---|---|---|---|
| Kyle | △ | △ | △ | △ | △ | | | |
| Rani | | | | | | | | |

**5.**

| Kyle has 3 more △ than Rani. | | | | | | | |
|---|---|---|---|---|---|---|---|
| Kyle | △ | △ | △ | △ | △ | | | |
| Rani | | | | | | | | |

Liz has 10 grapes.

◯ Circle who has more grapes. Underline who has fewer grapes.

◯ Fill in the table.

| 6. | Does Jon have more or fewer? | How many does Jon have? |
|---|---|---|
| (Liz) has 2 more grapes than <u>Jon</u>. | *fewer* | 10 (−) 2 |
| Liz has 2 fewer grapes than Jon. | | 10 ◯ 2 |
| Liz has 3 more grapes than Jon. | | 10 ◯ 3 |
| Liz has 3 fewer grapes than Jon. | | 10 ◯ 3 |

◯ Circle who has more. Underline who has fewer.

◯ Fill in the table.

| 7. | Does Sal have more or fewer? | How many does Sal have? |
|---|---|---|
| Amy has 8 grapes. <br> <u>Amy</u> has 3 fewer grapes than (Sal.) | *more* | 8 + 3 = 11 |
| Amy has 8 grapes. <br> Amy has 3 more grapes than Sal. | | _____ = ____ |
| Amy has 6 grapes. <br> Amy has 5 fewer grapes than Sal. | | _____ = ____ |
| Amy has 11 grapes. <br> Amy has 7 more grapes than Sal. | | _____ = ____ |

☐ Circle who has more pencils. Underline who has fewer pencils.
☐ Find how many pencils David has.

**8.**

Hanna has 26 pencils.

<u>Hanna</u> has 13 fewer pencils than (David.)

How many pencils does David have?

$$
\begin{array}{r}
  2\ 6 \\
  +\ 1\ 3 \\
  \hline
\end{array}
$$

**9.**

Hanna has 43 pencils.

(Hanna) has 22 more pencils than <u>David</u>.

How many pencils does David have?

**10.**

Hanna has 26 pencils.

Hanna has 37 fewer pencils than David.

How many pencils does David have?

**11.**

Hanna has 65 pencils.

Hanna has 18 more pencils than David.

How many pencils does David have?

**12.**

Hanna has 79 pencils.

Hanna has 43 more pencils than David.

How many pencils does David have?

☐ Circle groups of 10 tens.

☐ Draw the same number using hundreds and tens blocks.

**1.**

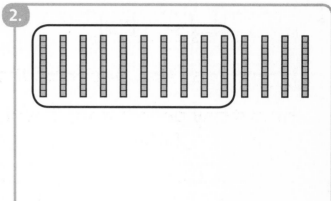

**2.**

**3.**

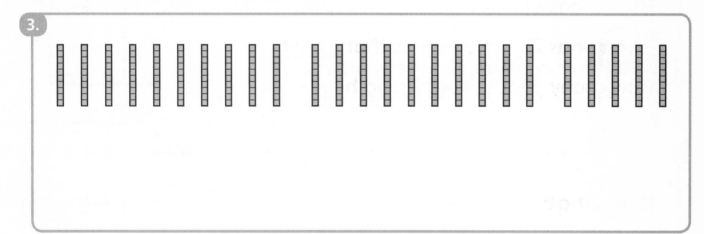

☐ What number do the hundreds blocks show?

**4.**

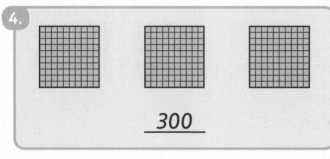

_300_

**5.**

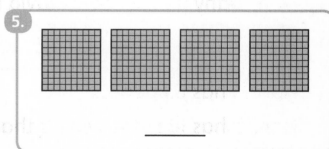

_____

**6.**

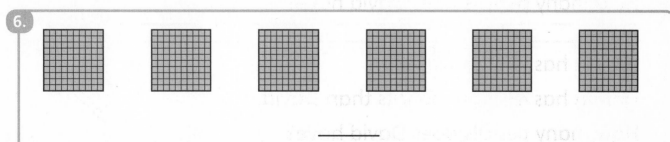

_____

Draw hundreds blocks to show the number.

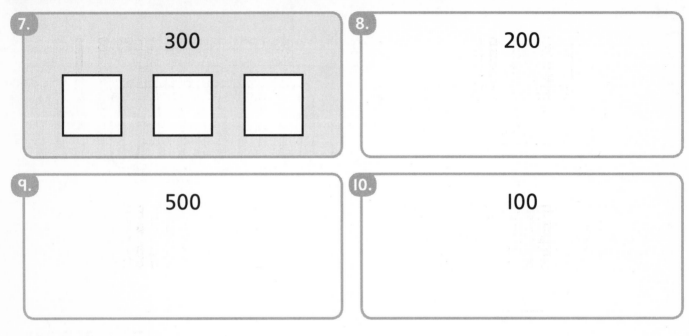

7. 300

8. 200

9. 500

10. 100

Write the number of hundreds blocks.
Write the number that the tens and ones blocks show.

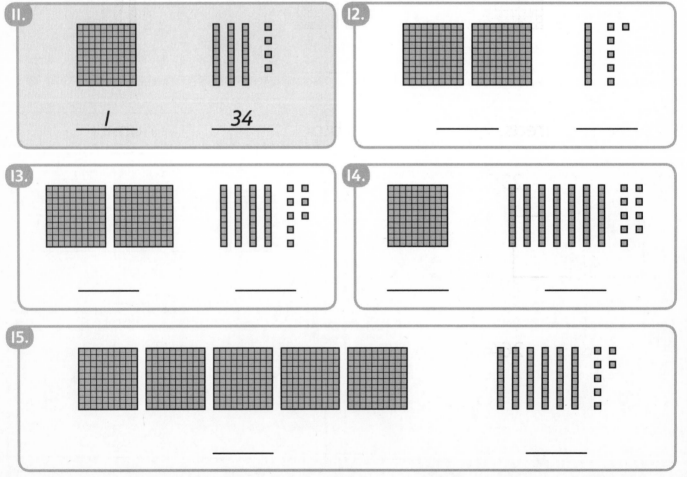

11. _1_  _34_

12. _____  _____

13. _____  _____

14. _____  _____

15. _____  _____

## ☐ What number do the blocks show?

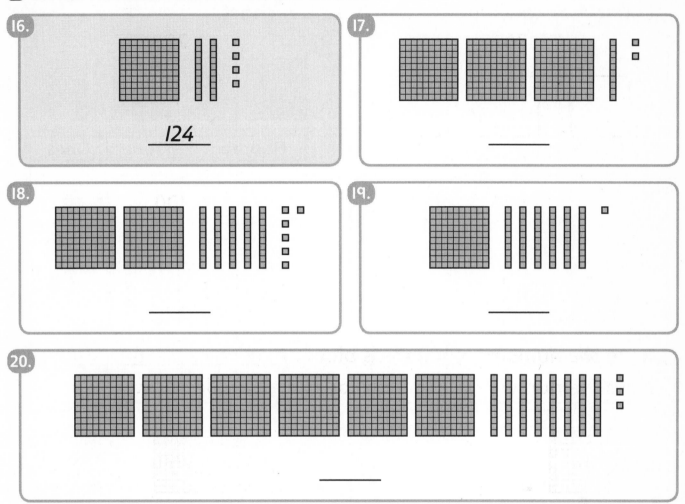

16. 124

17. _____

18. _____

19. _____

20. _____

## ☐ Draw hundreds, tens, and ones blocks to show the number.

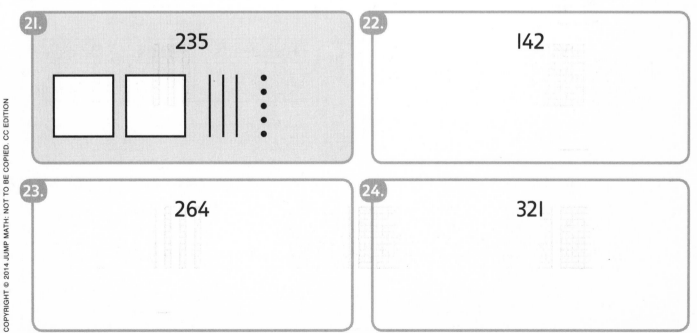

21. 235

22. 142

23. 264

24. 321

# NBT2-26 Place Value with 3 Digits (2)

☐ Count the blocks to fill in the **base ten chart**.

**1.**

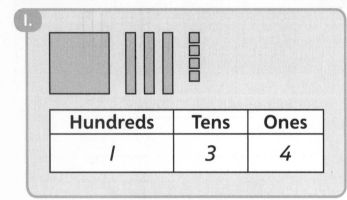

| Hundreds | Tens | Ones |
|----------|------|------|
| 1 | 3 | 4 |

**2.**

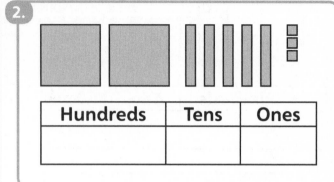

| Hundreds | Tens | Ones |
|----------|------|------|
| | | |

**3.**

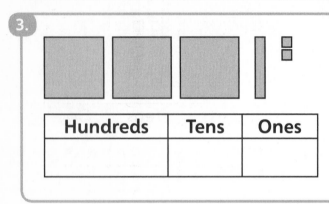

| Hundreds | Tens | Ones |
|----------|------|------|
| | | |

**4.**

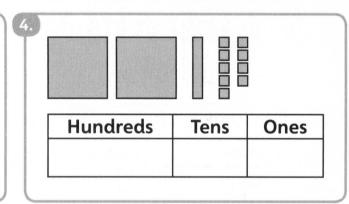

| Hundreds | Tens | Ones |
|----------|------|------|
| | | |

☐ Draw the number of blocks shown in the base ten chart.

**5.**

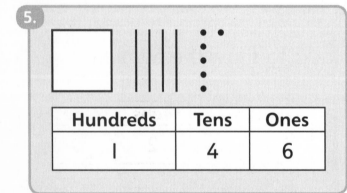

| Hundreds | Tens | Ones |
|----------|------|------|
| 1 | 4 | 6 |

**6.**

| Hundreds | Tens | Ones |
|----------|------|------|
| 2 | 5 | 3 |

**7.**

| Hundreds | Tens | Ones |
|----------|------|------|
| 2 | 3 | 7 |

**8.**

| Hundreds | Tens | Ones |
|----------|------|------|
| 1 | 2 | 4 |

☐ Count the blocks to fill in the base ten chart.

**9.**

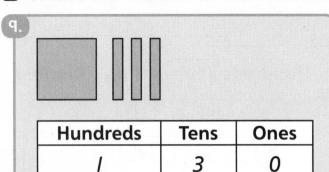

| Hundreds | Tens | Ones |
|----------|------|------|
| 1 | 3 | 0 |

**10.**

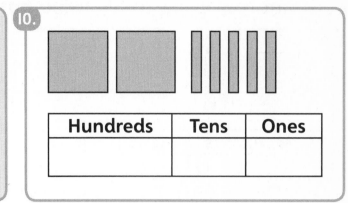

| Hundreds | Tens | Ones |
|----------|------|------|
|  |  |  |

**11.**

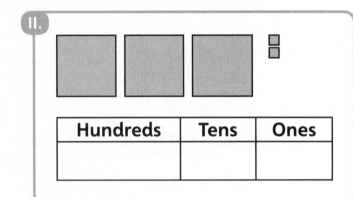

| Hundreds | Tens | Ones |
|----------|------|------|
|  |  |  |

**12.**

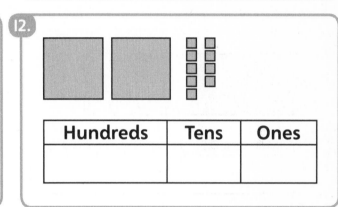

| Hundreds | Tens | Ones |
|----------|------|------|
|  |  |  |

☐ Draw the number of blocks shown in the base ten chart.

**13.**

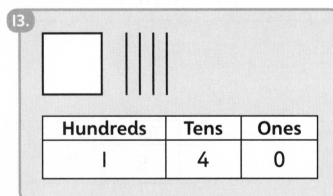

| Hundreds | Tens | Ones |
|----------|------|------|
| 1 | 4 | 0 |

**14.**

| Hundreds | Tens | Ones |
|----------|------|------|
| 3 | 2 | 0 |

**15.**

| Hundreds | Tens | Ones |
|----------|------|------|
| 2 | 0 | 7 |

**16.**

| Hundreds | Tens | Ones |
|----------|------|------|
| 1 | 0 | 4 |

☐ Count the hundreds, tens, and ones blocks.
☐ Write the number.

| | Hundreds | Tens | Ones | Number |
|---|---|---|---|---|
| | 1 | 4 | 5 | 145 |
| | | | | |
| | | | | |
| | | | | |
| | | | | |
| | | | | |

Fill in the table.

| 18. | | Hundreds | Tens | Ones | Number |
|---|---|---|---|---|---|
| | | 1 | 3 | 7 | 137 |
| | | | | | |
| | | 1 | 4 | 2 | |
| | | | | | 246 |
| | | 2 | 0 | 8 | |
| | | | | | 330 |
| | | 0 | 4 | 5 | |

# NBT2-27 Expanded Form

◯ Write what the digit 4 stands for.

| 1. 245 | 2. 431 | 3. 994 | 4. 734 | 5. 947 | 6. 463 |
|---|---|---|---|---|---|
| 40 | ____ | ____ | ____ | ____ | ____ |

◯ What does the underlined digit stand for?

| 7. 6̲53 | 8. 5̲37 | 9. 9̲78 | 10. 73̲4 | 11. 45̲2 | 12. 58̲9 |
|---|---|---|---|---|---|
| 600 | ____ | ____ | 30 | ____ | ____ |

| 13. 95̲1 | 14. 32̲6 | 15. 29̲2 | 16. 96̲8 | 17. 3̲72 | 18. 76̲1 |
|---|---|---|---|---|---|
| ____ | ____ | ____ | ____ | ____ | ____ |

| 19. BONUS | 20. BONUS | 21. BONUS |
|---|---|---|
| 2̲22 ____ | 22̲2 ____ | 222̲ ____ |

◯ Fill in the blanks for the base ten names.

22. 749 = ___7___ hundreds + ___4___ tens + ___9___ ones

23. 835 = _____ hundreds + _____ tens + _____ ones

24. 301 = _____ hundreds + _____ tens + _____ one

25. 120 = _____ hundred + _____ tens + _____ ones

☐ Fill in the blanks for the expanded form.

**26.** $749 = 700 + \underline{\phantom{ }40\phantom{ }} + 9$

**27.** $873 = 800 + \underline{\phantom{XXXX}} + 3$

**28.** $531 = \underline{\phantom{XXXX}} + 30 + 1$

**29.** $492 = \underline{\phantom{XXXX}} + 90 + 2$

**30.** $628 = 600 + 20 + \underline{\phantom{XXXX}}$

**31.** $341 = 300 + 40 + \underline{\phantom{XXXX}}$

**32.** $197 = \underline{\phantom{XXX}} + \underline{\phantom{XXX}} + \underline{\phantom{XXX}}$

**33.** $246 = \underline{\phantom{XXX}} + \underline{\phantom{XXX}} + \underline{\phantom{XXX}}$

**34.** $863 = \underline{\phantom{XXX}} + \underline{\phantom{XXX}} + \underline{\phantom{XXX}}$

**35.** $752 = \underline{\phantom{XXX}} + \underline{\phantom{XXX}} + \underline{\phantom{XXX}}$

☐ Write the total.

**36.** $600 + 50 + 3 = \underline{\phantom{ }653\phantom{ }}$

**37.** $900 + 70 + 5 = \underline{\phantom{XXXX}}$

**38.** $500 + 90 + 7 = \underline{\phantom{XXXX}}$

**39.** $200 + 80 + 2 = \underline{\phantom{XXXX}}$

**40.** $300 + 20 + 4 = \underline{\phantom{XXXX}}$

**41.** $100 + 40 + 6 = \underline{\phantom{XXXX}}$

**42.** $800 + 10 + 9 = \underline{\phantom{XXXX}}$

**43.** $700 + 30 + 1 = \underline{\phantom{XXXX}}$

⬜ Write the total.

**44.**
2 hundreds + 4 tens + 5 ones

_245_

**45.**
4 hundreds + 8 tens + 7 ones

_____

**46.**
6 hundreds + 3 tens + 1 one

_____

**47.**
8 hundreds + 1 ten + 8 ones

_____

⬜ Write the total.

**48.**
$$600 + 50 + 3 = \boxed{653}$$

**49.**
$$600 + 50$$
☐

**50.**
$$600 + 3$$
☐

**51.**
$$800 + 20 + 7$$
☐

**52.**
$$800 + 20$$
☐

**53.**
$$800 + 7$$
☐

**54.** 200 + 40 = _____

**55.** 300 + 60 = _____

**56.** 700 + 10 = _____

**57.** 200 + 4 = _____

**58.** 300 + 6 = _____

**59.** 700 + 1 = _____

**60.** 400 + 50 = _____

**61.** 500 + 7 = _____

**62.** 900 + 3 = _____

**63.** 100 + 80 = _____

**64.** 600 + 5 = _____

**65.** 400 + 70 = _____

**66.**
Ben says 600 + 5 = 650. Explain his mistake.

Number words for the ones:

| zero | one | two | three | four |
|------|-----|-----|-------|------|
| five | six | seven | eight | nine |

☐ Write the number word.

**1.**
8  _eight_

**2.**
6  _____

**3.**
7  _____

**4.**
1  _____

**5.**
5  _____

**6.**
9  _____

**7.**
0  _____

**8.**
4  _____

**9.**
3  _____

**10.**
2  _____

**11.**
3  _____

**12.**
8  _____

☐ Write the endings for the number words.

**13.**
16  six _teen_
60  six _ty_

**14.**
15  fif_____
50  fif_____

**15.**
18  eigh_____
80  eigh_____

**16.**
19  nine_____
90  nine_____

**17.**
14  four_____
40  for_____

**18.**
17  seven_____
70  seven_____

**19.**
13  thir_____
30  thir_____

**20. BONUS**
60  six_____
16  six_____

**21. BONUS**
50  fif_____
15  fif_____

☐ Write the number word. Use the words in the box.

| ten   eleven   twelve   thirteen   fifteen   twenty |

**22.**
10 __ten__

**23.**
13 _____

**24.**
12 _____

**25.**
20 _____

**26.**
11 _____

**27.**
15 _____

Number words for multiples of 10:

ten          twenty          thirty          forty          fifty

sixty          seventy          eighty          ninety

☐ Write the number word for the total. Use the addition to help.

**28.**
$63 = 60 + 3$

__sixty__ - __three__

**29.**
$48 = 40 + 8$

_____ - __eight__

**30.**
$95 = 90 + 5$

_____ - _____

**31.**
$72 = 70 + 2$

_____ - _____

**32.**
$21 = 20 + 1$

_____ - _____

**33.**
$56 = 50 + 6$

_____ - _____

**34.**
$39 = 30 + 9$

_____ - _____

**35.**
$87 = 80 + 7$

_____ - _____

**36.**
$69 = 60 + 9$

_____ - _____

COPYRIGHT © 2014 JUMP MATH: NOT TO BE COPIED. CC EDITION

Number and Operations in Base Ten 2-28

## ☐ Write the numbers.

**37.**
fifteen    _15_

fifty     _50_

**38.**
sixteen    _____

sixty      _____

**39.**
eighteen    _____

eighty      _____

**40.**
ninety    _____

nineteen   _____

**41.**
seventy    _____

seventeen   _____

**42.**
thirteen    _____

thirty      _____

**43.**
fourteen    _____

**44.**
fifteen    _____

**45.**
forty    _____

## ☐ Write the number.

**46.**
forty-seven

_47_

**47.**
eighty-five

_____

**48.**
ninety-four

_____

**49.**
twenty-six

_____

**50.**
thirty-two

_____

**51.**
seventy-three

_____

**52.**
fifty-one

_____

**53.**
sixty-eight

_____

**54.**
zero

_____

**55.**
eleven

_____

**56.**
ten

_____

**57.**
twelve

_____

Number words for the ones:

| zero | one | two | three | four |
|------|-----|-----|-------|------|
| five | six | seven | eight | nine |

Number words for 10 to 19:

| ten | eleven | twelve | thirteen | fourteen |
|-----|--------|--------|----------|----------|
| fifteen | sixteen | seventeen | eighteen | nineteen |

☐ Write the number word.

**1.**
800

_eight hundred_

**2.**
500

_____

**3.**
300

_____

**4.**
200

_____

☐ Underline the hundreds digit.
☐ What does the hundreds digit stand for?

**5.**
<u>7</u>69

_seven hundred_

**6.**
942

_____

**7.**
678

_____

**8.**
427

_____

**9.**
186

_____

**10.**
841

_____

◻ Circle the last 2 digits.
◻ Write the number word for the last 2 digits.

**11.**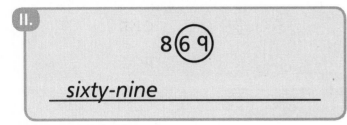
8 ⑥ 9
_sixty-nine_____

**12.**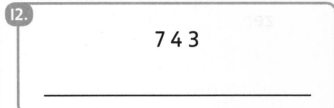
7 4 3
_____

**13.**
5 7 2
_____

**14.**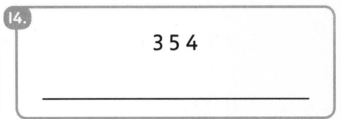
3 5 4
_____

**15.**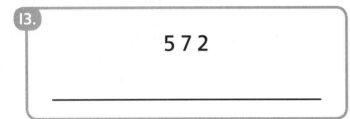
9 8 1
_____

**16.**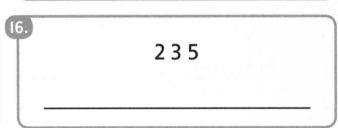
2 3 5
_____

◻ Write what the hundreds digit stands for.
◻ Write the rest of the number word.

**17.**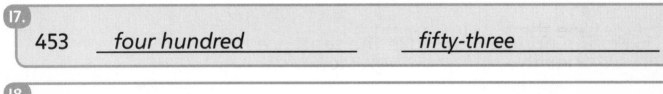
453   _four hundred_____   _fifty-three_____

**18.**
689   _____   _eighty-nine_____

**19.**
547   _____   _____

**20.**
228   _____   _____

**21.**
316   _____   _____

◯ Circle the larger number.

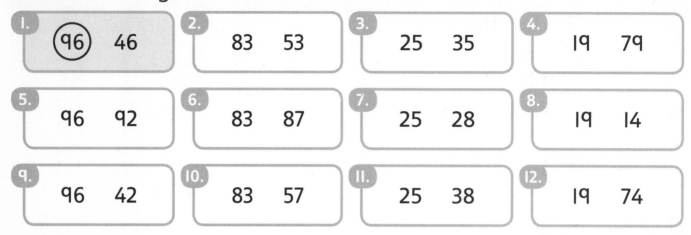

1. **96** 46
2. 83 53
3. 25 35
4. 19 79
5. 96 92
6. 83 87
7. 25 28
8. 19 14
9. 96 42
10. 83 57
11. 25 38
12. 19 74

◯ Circle **Yes** or **No**.

13. 19 is **greater** than 15
**Yes** No

14. 28 is **less** than 36
Yes No

15. 62 is less than 59
Yes No

16. 83 is greater than 91
Yes No

◯ Fill in the numbers to make the sentence true.

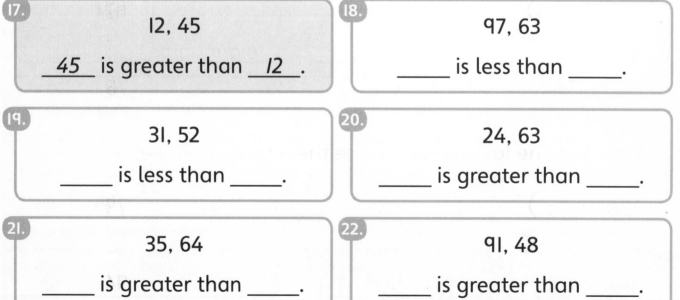

17. 12, 45
_45_ is greater than _12_.

18. 97, 63
_____ is less than _____.

19. 31, 52
_____ is less than _____.

20. 24, 63
_____ is greater than _____.

21. 35, 64
_____ is greater than _____.

22. 91, 48
_____ is greater than _____.

| 3 < 5 | 6 > 2 |
|---|---|
| 3 is less than 5 | 6 is greater than 2 |

☐ Circle **Yes** or **No**.

**23.** 11 > 32
Yes (No)

**24.** 78 > 36
Yes No

**25.** 37 < 53
Yes No

**26.** 34 < 15
Yes No

☐ Write > or <.

**27.** 75 (>) 35

**28.** 62 ◯ 42

**29.** 27 ◯ 29

**30.** 19 ◯ 15

**31.** 94 ◯ 52

**32.** 22 ◯ 63

**33.** 56 ◯ 74

**34.** 48 ◯ 81

**35.** 37 ◯ 68

**36.** 83 ◯ 59

**37.** 71 ◯ 27

**38.** 84 ◯ 92

☐ Underline the hundreds digit. Circle the greater number.

**39.** <u>1</u>58 (<u>3</u>58)

**40.** <u>8</u>75 <u>2</u>75

**41.** 953 453

**42.** 622 722

**43.** 158 342

**44.** 579 643

**45.** 491 254

**46.** 763 507

☐ Underline the last 2 digits. Circle the greater number.

**47.** 1<u>58</u> (1<u>64</u>)

**48.** 5<u>75</u> 5<u>23</u>

**49.** 814 883

**50.** 794 777

**51.** 253 231

**52.** 643 608

**53.** 312 319

**54.** 946 981

☐ Underline the first digit that is different. Look at the hundreds digit first.

☐ Circle the greater number.

| 55. | 56. | 57. | 58. | 59. | 60. |
|---|---|---|---|---|---|
| <u>1</u>58 ⃝463 | 358 165 | 875 342 | 9<u>5</u>3 9<u>3</u>2 | 453 487 | 622 693 |

| 61. | 62. | 63. | 64. | 65. | 66. |
|---|---|---|---|---|---|
| 15<u>8</u> 15<u>2</u> | 342 347 | 579 572 | 491 298 | 254 352 | 507 309 |

☐ Circle the correct sign.

| 67. | 68. | 69. | 70. |
|---|---|---|---|
| 135 &lt; &gt; 235 | 657 &lt; &gt; 457 | 923 &lt; &gt; 723 | 419 &lt; &gt; 819 |

| 71. | 72. | 73. | 74. |
|---|---|---|---|
| 167 &lt; &gt; 189 | 359 &lt; &gt; 356 | 784 &lt; &gt; 794 | 523 &lt; &gt; 506 |

| 75. | 76. | 77. | 78. |
|---|---|---|---|
| 146 &lt; &gt; 169 | 157 &lt; &gt; 265 | 743 &lt; &gt; 562 | 989 &lt; &gt; 691 |

| 79. | 80. | 81. | 82. |
|---|---|---|---|
| 325 &lt; &gt; 327 | 452 &lt; &gt; 412 | 671 &lt; &gt; 573 | 812 &lt; &gt; 259 |

| 83. | 84. | 85. | 86. |
|---|---|---|---|
| 479 &lt; = &gt; 555 | 123 &lt; = &gt; 123 | 865 &lt; = &gt; 567 | 753 &lt; = &gt; 753 |

# NBT2-31 Adding 3-Digit Numbers

☐ Add by counting on.

**1.**  21 + 5 = \_\_\_\_\_

**2.**  91 + 7 = \_\_\_\_\_

**3.**  85 + 2 = \_\_\_\_\_

**4.**  121 + 5 = \_\_\_\_\_

**5.**  191 + 7 = \_\_\_\_\_

**6.**  285 + 2 = \_\_\_\_\_

**7.**  35 + 6 = \_\_\_\_\_

**8.**  74 + 8 = \_\_\_\_\_

**9.**  69 + 5 = \_\_\_\_\_

**10.**  135 + 6 = \_\_\_\_\_

**11.**  274 + 8 = \_\_\_\_\_

**12.**  769 + 5 = \_\_\_\_\_

☐ Write the number of hundreds, tens, and ones.
☐ Add.

**13.**

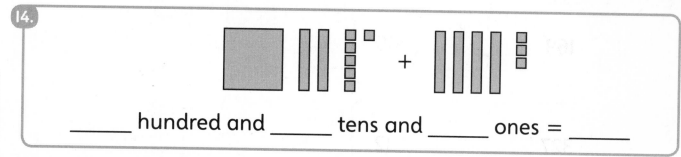

\_\_\_*1*\_\_\_ hundred and \_\_\_*5*\_\_\_ tens and \_\_\_*6*\_\_\_ ones = \_\_\_*156*\_\_\_

**14.**

_____ hundred and _____ tens and _____ ones = _____

**15.**

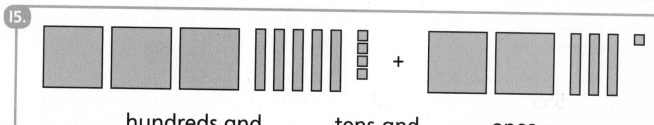

_____ hundreds and _____ tens and _____ ones = _____

☐ Write the numbers that the blocks show.
☐ Use the blocks to help you add.

**16.**

_135_ + _21_ = _156_

**17.**

_____ + _____ = _____

**18.**

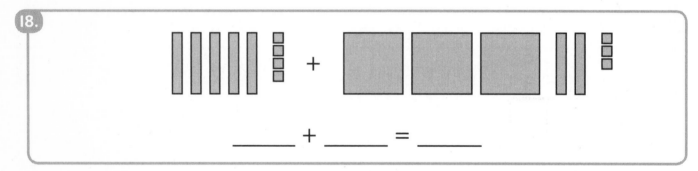

_____ + _____ = _____

**19.**

_____ + _____ = _____

**20.**

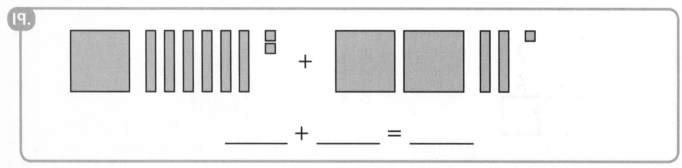

_____ + _____ = _____

　　　　　　　　**Number and Operations in Base Ten 2-31**

Draw hundreds, tens, and ones to show the numbers.
Use the picture to help you add.

**21.**

$$\begin{array}{r} 2\ 3\ 5 \\ +\ 1\ 4\ 3 \\ \hline 3\ 7\ 8 \end{array}$$

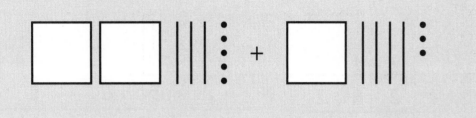

**22.**

$$\begin{array}{r} 1\ 2\ 4 \\ +\ 2\ 3\ 1 \\ \hline \phantom{000} \end{array}$$

**23.**

$$\begin{array}{r} 3\ 2\ 6 \\ +\ 1\ 2\ 3 \\ \hline \phantom{000} \end{array}$$

**24.**

$$\begin{array}{r} 1\ 5\ 2 \\ +\ 1\ 2\ 4 \\ \hline \phantom{000} \end{array}$$

**25. BONUS**

The mother panda eats 243 pounds of bamboo a week. The father panda eats 254 pounds of bamboo a week. How much bamboo do they eat altogether?

# NBT2-32 Addition Using Expanded Form

☐ Write the numbers using base ten names.

☐ Add.

**1.**

$$
\begin{array}{r}
4\ 3\ 5 \\
+\ 2\ 6\ 2 \\
\hline
6\ 9\ 7
\end{array}
$$

<u> 4 </u> hundreds + <u> 3 </u> tens + <u> 5 </u> ones

+ <u> 2 </u> hundreds + <u> 6 </u> tens + <u> 2 </u> ones

<u> 6 </u> hundreds + <u> 9 </u> tens + <u> 7 </u> ones

**2.**

$$
\begin{array}{r}
5\ 4\ 7 \\
+\ 4\ 3\ 1 \\
\hline
\end{array}
$$

<u> 5 </u> hundreds + <u> 4 </u> tens + <u> 7 </u> ones

+ <u> 4 </u> hundreds + <u> 3 </u> tens + <u> 1 </u> one

___ hundreds + ___ tens + ___ ones

**3.**

$$
\begin{array}{r}
4\ 1\ 3 \\
+\ 2\ 5\ 5 \\
\hline
\end{array}
$$

___ hundreds + ___ ten + ___ ones

+ ___ hundreds + ___ tens + ___ ones

___ hundreds + ___ tens + ___ ones

**4.**

$$
\begin{array}{r}
6\ 3\ 7 \\
+\ 1\ 5\ 1 \\
\hline
\end{array}
$$

___ hundreds + ___ tens + ___ ones

+ ___ hundred + ___ tens + ___ one

___ hundreds + ___ tens + ___ ones

**5.**

$$
\begin{array}{r}
6\ 5\ 8 \\
+\ 2\ 3\ 1 \\
\hline
\end{array}
$$

___ hundreds + ___ tens + ___ ones

+ ___ hundreds + ___ tens + ___ one

___ hundreds + ___ tens + ___ ones

☐ Write the numbers in expanded form.

☐ Add.

**6.**

$$
\begin{array}{r}
4\ 3\ 5 \\
+\ 2\ 6\ 2 \\
\hline
6\ 9\ 7
\end{array}
$$

$$\underline{\ 400\ }\ +\ \underline{\ 30\ }\ +\ \underline{\ 5\ }$$
$$+\ \underline{\ 200\ }\ +\ \underline{\ 60\ }\ +\ \underline{\ 2\ }$$
$$\underline{\ 600\ }\ +\ \underline{\ 90\ }\ +\ \underline{\ 7\ }$$

**7.**

$$
\begin{array}{r}
3\ 5\ 6 \\
+\ 4\ 4\ 2 \\
\hline
\phantom{000}
\end{array}
$$

$$\underline{\ 300\ }\ +\ \underline{\ 50\ }\ +\ \underline{\ 6\ }$$
$$+\ \underline{\quad}\ +\ \underline{\quad}\ +\ \underline{\quad}$$
$$\underline{\quad}\ +\ \underline{\quad}\ +\ \underline{\quad}$$

**8.**

$$
\begin{array}{r}
3\ 4\ 3 \\
+\ 1\ 4\ 5 \\
\hline
\phantom{000}
\end{array}
$$

$$\underline{\quad}\ +\ \underline{\quad}\ +\ \underline{\quad}$$
$$+\ \underline{\quad}\ +\ \underline{\quad}\ +\ \underline{\quad}$$
$$\underline{\quad}\ +\ \underline{\quad}\ +\ \underline{\quad}$$

**9.**

$$
\begin{array}{r}
1\ 3\ 7 \\
+\ 8\ 3\ 2 \\
\hline
\phantom{000}
\end{array}
$$

$$\underline{\quad}\ +\ \underline{\quad}\ +\ \underline{\quad}$$
$$+\ \underline{\quad}\ +\ \underline{\quad}\ +\ \underline{\quad}$$
$$\underline{\quad}\ +\ \underline{\quad}\ +\ \underline{\quad}$$

**10. BONUS**

$$
\begin{array}{r}
2\ 0\ 1 \\
+\ 3\ 6\ 1 \\
\hline
\phantom{000}
\end{array}
$$

$$\underline{\quad}\ +\ \underline{\quad}\ +\ \underline{\quad}$$
$$+\ \underline{\quad}\ +\ \underline{\quad}\ +\ \underline{\quad}$$
$$\underline{\quad}\ +\ \underline{\quad}\ +\ \underline{\quad}$$

☐ Write the numbers using base ten names.
☐ Add.

**11.**

```
    4   0   5
 +  1   6   2
 ───────────
    5   6   7
```

__4__ hundreds + __0__ tens + __5__ ones
+ __1__ hundred + __6__ tens + __2__ ones
───────────────────────────────────────
__5__ hundreds + __6__ tens + __7__ ones

**12.**

```
    5   4   7
 +  4   0   1
 ───────────
```

__5__ hundreds + __4__ tens + __7__ ones
+ ___ hundreds + ___ tens + ___ one
───────────────────────────────────────
___ hundreds + ___ tens + ___ ones

**13.**

```
    7   1   0
 +  2   5   5
 ───────────
```

___ hundreds + ___ ten + ___ ones
+ ___ hundreds + ___ tens + ___ ones
───────────────────────────────────────
___ hundreds + ___ tens + ___ ones

**14.**

```
    7   5   8
 +      3   1
 ───────────
```

___ hundreds + ___ tens + ___ ones
+ ___ hundreds + ___ tens + ___ one
───────────────────────────────────────
___ hundreds + ___ tens + ___ ones

**15.**

```
    8   2   0
 +  1   5   0
 ───────────
```

___ hundreds + ___ tens + ___ ones
+ ___ hundred + ___ tens + ___ ones
───────────────────────────────────────
___ hundreds + ___ tens + ___ ones

Number and Operations in Base Ten 2-32

☐ Use the blocks to add.

**1.**

```
    3  4  5
 +  2  3  4
 ─────────
    5  7  9
```

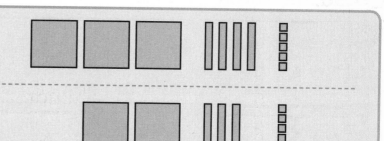

**2.**

```
    3  2  4
 +  3  5  1
 ─────────
```

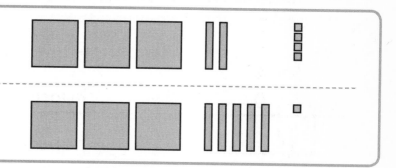

**3.**

```
    2  7  5
 +  4  1  4
 ─────────
```

**4.**

```
    3  0  6
 +  4  7  1
 ─────────
```

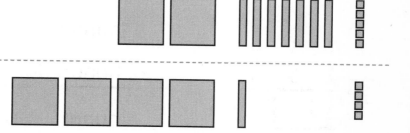

**5.**

```
    5  5  3
 +     2  5
 ─────────
```

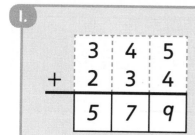
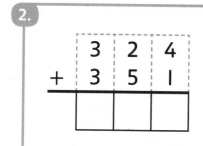
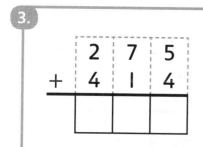
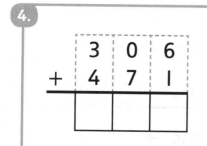

☐ Draw hundreds, tens, and ones to show the addition.
☐ Use the picture to add.

**6.**

|   | 2 | 4 | 3 |
|---|---|---|---|
| + | 3 | 2 | 5 |
|   | 5 | 6 | 8 |

**7.**

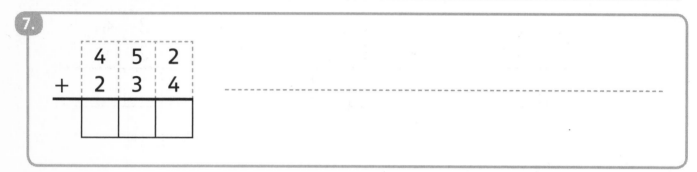

|   | 4 | 5 | 2 |
|---|---|---|---|
| + | 2 | 3 | 4 |
|   |   |   |   |

**8.**

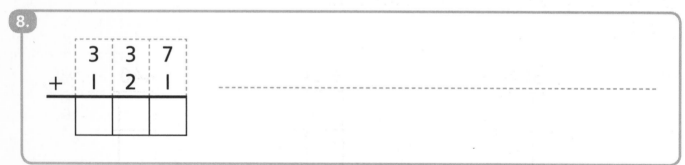

|   | 3 | 3 | 7 |
|---|---|---|---|
| + | 1 | 2 | 1 |
|   |   |   |   |

**9.**

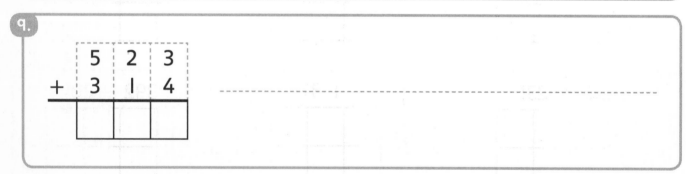

|   | 5 | 2 | 3 |
|---|---|---|---|
| + | 3 | 1 | 4 |
|   |   |   |   |

**10.**

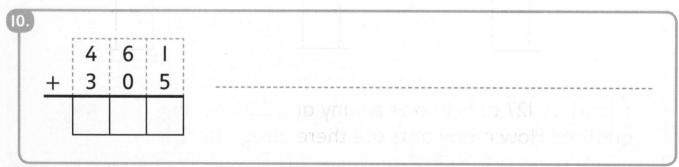

|   | 4 | 6 | 1 |
|---|---|---|---|
| + | 3 | 0 | 5 |
|   |   |   |   |

Number and Operations in Base Ten 2-33

☐ Add the ones. Add the tens. Add the hundreds.

**11.**
```
    2  5  3
+   4  2  6
  ─────────
    6  7  9
```

**12.**
```
    7  3  1
+   1  5  8
  ─────────
```

**13.**
```
    6  4  1
+   1  3  2
  ─────────
```

**14.**
```
    3  2  0
+   1  7  4
  ─────────
```

**15.**
```
    4  8  1
+   5  0  6
  ─────────
```

**16.**
```
    2  4  5
+   2  1  3
  ─────────
```

☐ Use the grid to write the addition.

☐ Add.

**17.** 325 + 304
```
    3  2  5
+   3  0  4
  ─────────
    6  2  9
```

**18.** 612 + 346

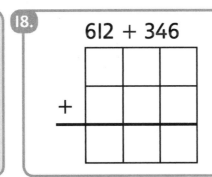

**19.** 440 + 249

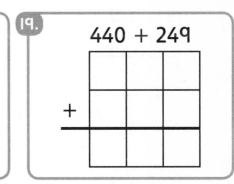

**20.** 701 + 231

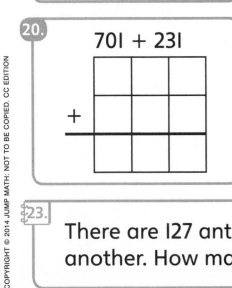

**21.** 216 + 370

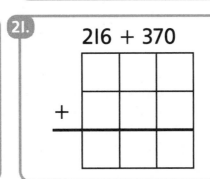

**22.** 392 + 201

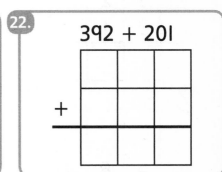

**23.**

There are 127 ants in one colony and 231 ants in another. How many ants are there altogether?

# NBT2-34 Using Base Ten Blocks to Add (Regrouping)

☐ Use the blocks to add.
☐ Write I to show regrouping.

1.

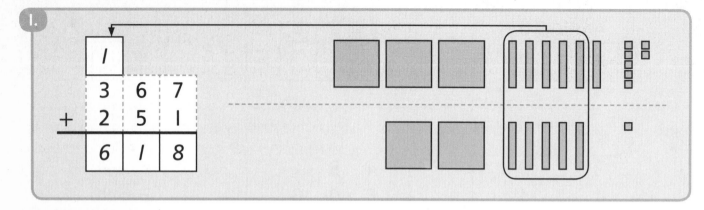

|   | I |   |   |
|---|---|---|---|
|   | 3 | 6 | 7 |
| + | 2 | 5 | 1 |
|   | 6 | I | 8 |

2.

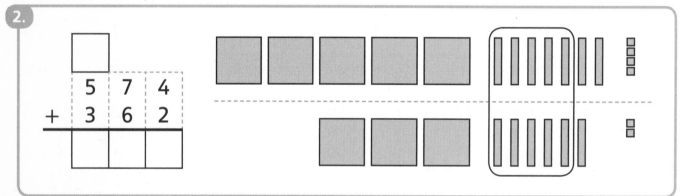

|   |   |   |   |
|---|---|---|---|
|   | 5 | 7 | 4 |
| + | 3 | 6 | 2 |
|   |   |   |   |

3.

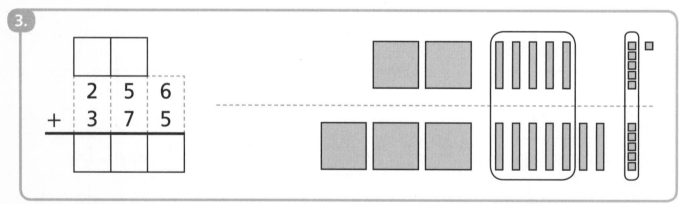

|   |   |   |   |
|---|---|---|---|
|   | 2 | 5 | 6 |
| + | 3 | 7 | 5 |
|   |   |   |   |

4. BONUS

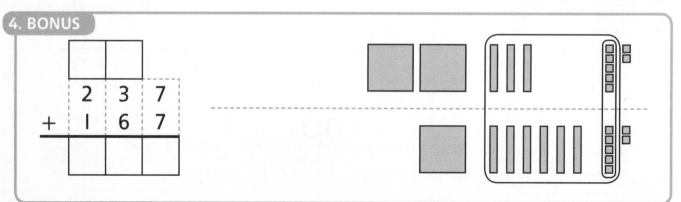

|   |   |   |   |
|---|---|---|---|
|   | 2 | 3 | 7 |
| + | 1 | 6 | 7 |
|   |   |   |   |

 Regroup the tens.

○ Add.

**5.**

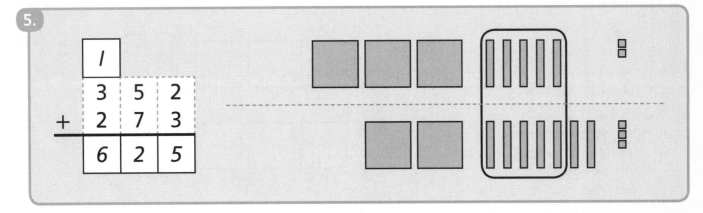

| 1 | | |
|---|---|---|
| 3 | 5 | 2 |
| + 2 | 7 | 3 |
| 6 | 2 | 5 |

**6.**

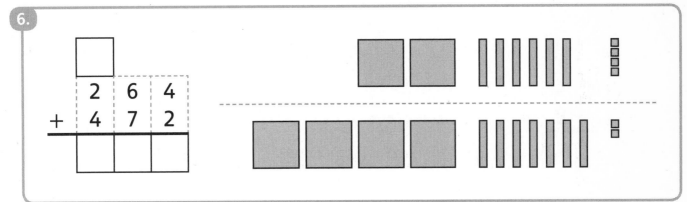

| | | |
|---|---|---|
| 2 | 6 | 4 |
| + 4 | 7 | 2 |
| | | |

**7.**

| | | |
|---|---|---|
| | 6 | 5 |
| + | 8 | 3 |
| | | |

**8. BONUS**

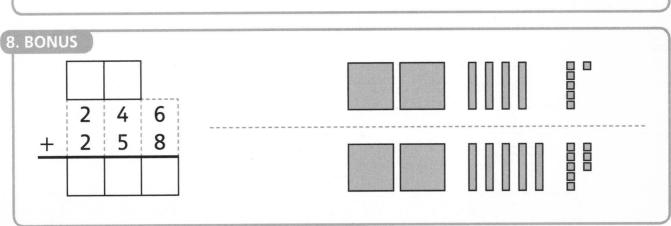

| | | |
|---|---|---|
| 2 | 4 | 6 |
| + 2 | 5 | 8 |
| | | |

☐ Draw hundreds, tens, and ones. Show the regrouping.
☐ Add.

**9.**

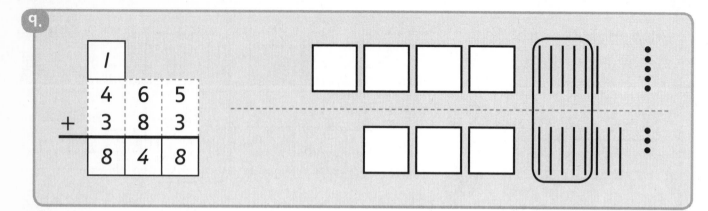

| | 1 | | |
|---|---|---|---|
| | 4 | 6 | 5 |
| + | 3 | 8 | 3 |
| | 8 | 4 | 8 |

**10.**

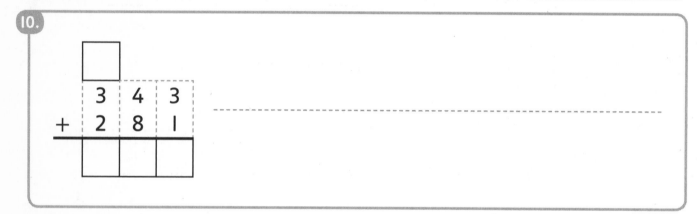

| | | | |
|---|---|---|---|
| | 3 | 4 | 3 |
| + | 2 | 8 | 1 |
| | | | |

**11.**

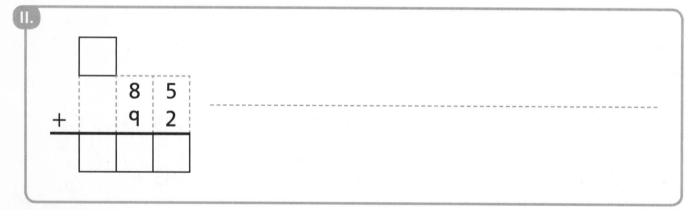

| | | | |
|---|---|---|---|
| | | 8 | 5 |
| + | | 9 | 2 |
| | | | |

**12. BONUS**

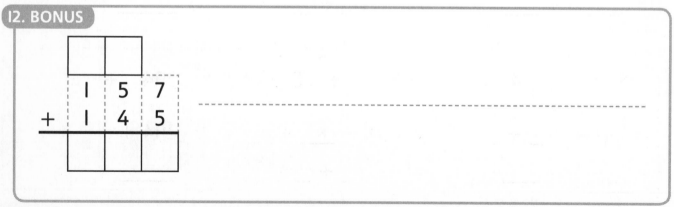

| | | | |
|---|---|---|---|
| | 1 | 5 | 7 |
| + | 1 | 4 | 5 |
| | | | |

**Number and Operations in Base Ten 2-34**

# NBT2-35 Using Strategies to Add

☐ Write the missing number.

**1.** 498 + __2__ = 500

**2.** 397 + ____ = 400

**3.** 692 + ____ = 700

**4.** 791 + ____ = 800

**5.** 293 + ____ = 300

**6.** 194 + ____ = 200

**7.** 595 + ____ = 600

**8.** 899 + ____ = 900

**9.** 496 + ____ = 500

☐ Fill in the blanks so that the first two numbers add to a multiple of 100.

☐ Add to make the next multiple of 100.

**10.** 498 + 7

= 498 + __2__ + __5__

= 500 + __5__

**11.** 393 + 9

= 393 + __7__ + ____

= 400 + ____

**12.** 697 + 5

= 697 + __3__ + ____

= 700 + ____

**13.** 396 + 7

= 396 + ____ + ____

= ____ + ____

**14.** 798 + 6

= 798 + ____ + ____

= ____ + ____

**15.** 495 + 9

= 495 + ____ + ____

= ____ + ____

**16.** 597 + 4

= 597 + ____ + ____

= ____ + ____

**17.** 899 + 8

= 899 + ____ + ____

= ____ + ____

**18.** 294 + 8

= 294 + ____ + ____

= ____ + ____

○ Fill in the blanks so that the first two numbers add to a multiple of 100.

○ Add to make the next multiple of 100.

○ Add.

**19.**

396 + 7

= 396 + _4_ + _3_

= _400_ + _3_

= _403_

**20.**

798 + 6

= 798 + ____ + ____

= _800_ + ____

= ____

**21.**

597 + 4

= 597 + ____ + ____

= ____ + ____

= ____

**22.**

899 + 8

= 899 + ____ + ____

= ____ + ____

= ____

**23.**

193 + 9

= 193 + ____ + ____

= ____ + ____

= ____

**24.**

697 + 4

= 697 + ____ + ____

= ____ + ____

= ____

**25. BONUS**

597 + 14

= 597 + _3_ + _11_

= ____ + ____

= ____

**26. BONUS**

899 + 58

= 899 + ____ + ____

= ____ + ____

= ____

◯ Add.

**27.**

$793 + 8 = \underline{\quad793\quad} + \underline{\ 7\ } + \underline{\ 1\ }$

$\phantom{793 + 8 } = \underline{\ 800\ } + \underline{\ 1\ }$

$\phantom{793 + 8 } = \underline{\ 801\ }$

**28.**

$897 + 6 = \underline{\qquad} + \underline{\quad} + \underline{\quad}$

$\phantom{897 + 6 } = \underline{\qquad} + \underline{\qquad}$

$\phantom{897 + 6 } = \underline{\qquad}$

**29.**

$596 + 36 = \underline{\qquad} + \underline{\quad} + \underline{\quad}$

$\phantom{596 + 36 } = \underline{\qquad} + \underline{\qquad}$

$\phantom{596 + 36 } = \underline{\qquad}$

**30.**

$498 + 57 = \underline{\qquad} + \underline{\quad} + \underline{\quad}$

$\phantom{498 + 57 } = \underline{\qquad} + \underline{\qquad}$

$\phantom{498 + 57 } = \underline{\qquad}$

**31.** $495 + 9$

**32.** $294 + 8$

**33.** $396 + 6$

**34.** $694 + 7$

**35.** $297 + 36$

**36.** $398 + 25$

**37.** $797 + 66$

**38.** $893 + 99$

**39. BONUS**

Anna wants to collect 500 stamps. She has 496 stamps in her book. Peter gives her 7 more stamps. How many more than 500 stamps does Anna have now?

**40. BONUS**

Paul says that $395 + 8$ is the same as $400 + 3$.
Is he correct? Explain.

# NBT2-36 Using Place Value to Subtract (No Regrouping)

☐ Cross out blocks to show the subtraction.
☐ Write how many blocks are left.

1.

|   | 2 | 3 | 5 |
|---|---|---|---|
| − | 1 | 2 | 3 |
|   | 1 | 1 | 2 |

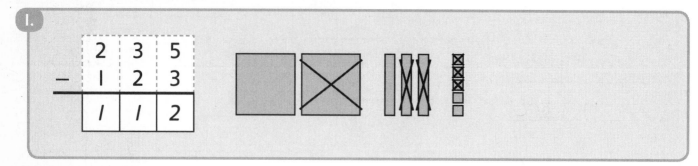

2.

|   | 3 | 4 | 6 |
|---|---|---|---|
| − | 1 | 3 | 2 |
|   |   |   |   |

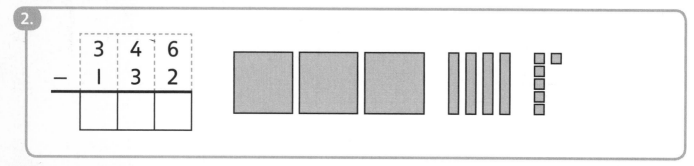

3.

|   | 4 | 6 | 7 |
|---|---|---|---|
| − | 1 | 4 | 5 |
|   |   |   |   |

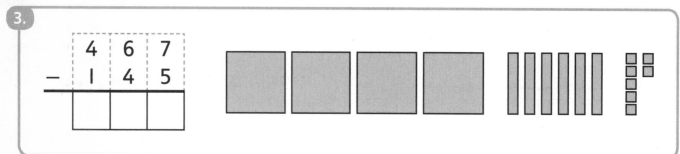

4.

|   | 3 | 5 | 2 |
|---|---|---|---|
| − | 2 | 2 | 1 |
|   |   |   |   |

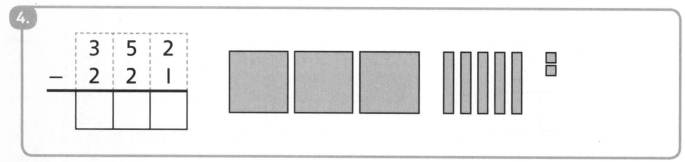

5.

|   | 4 | 0 | 5 |
|---|---|---|---|
| − | 1 | 0 | 3 |
|   |   |   |   |

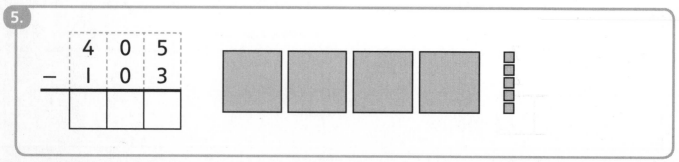

Number and Operations in Base Ten 2-36

☐ Draw a picture to show the subtraction.
☐ Subtract.

**6.**

$$
\begin{array}{r}
3\ 4\ 7 \\
-\ 2\ 1\ 3 \\
\hline
1\ 3\ 4
\end{array}
$$

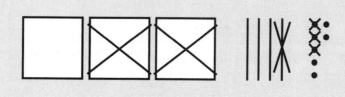

**7.**

$$
\begin{array}{r}
2\ 4\ 6 \\
-\ 1\ 2\ 5 \\
\hline
\phantom{0}\ \phantom{0}\ \phantom{0}
\end{array}
$$

**8.**

$$
\begin{array}{r}
5\ 2\ 3 \\
-\ 3\ 0\ 2 \\
\hline
\phantom{0}\ \phantom{0}\ \phantom{0}
\end{array}
$$

**9. BONUS**

$$
\begin{array}{r}
4\ 5\ 0 \\
-\ 2\ 1\ 0 \\
\hline
\phantom{0}\ \phantom{0}\ \phantom{0}
\end{array}
$$

**10. BONUS**

$$
\begin{array}{r}
5\ 0\ 7 \\
-\ 3\ 0\ 4 \\
\hline
\phantom{0}\ \phantom{0}\ \phantom{0}
\end{array}
$$

○ Subtract the ones. Subtract the tens. Subtract the hundreds.

**11.**
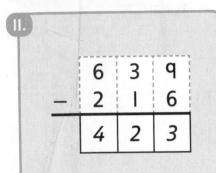
```
    6  3  9
 -  2  1  6
 ─────────
    4  2  3
```

**12.**
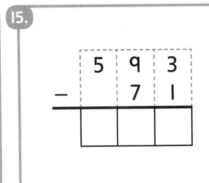
```
    7  5  6
 -  3  4  3
 ─────────
```

**13.**
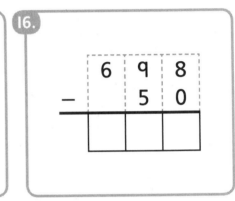
```
    5  8  9
 -  2  3  5
 ─────────
```

**14.**
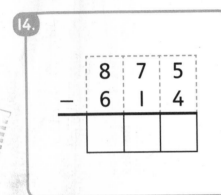
```
    8  7  5
 -  6  1  4
 ─────────
```

**15.**
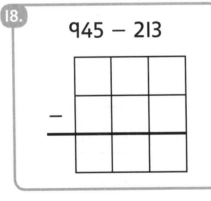
```
    5  9  3
 -     7  1
 ─────────
```

**16.**
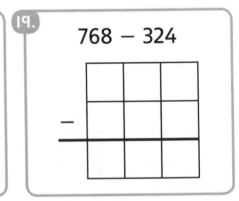
```
    6  9  8
 -     5  0
 ─────────
```

○ Write the subtraction in the grid.
○ Subtract.

**17.** 879 − 526
```
    8  7  9
 -  5  2  6
 ─────────
    3  5  3
```

**18.** 945 − 213
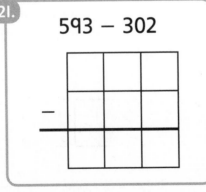

**19.** 768 − 324
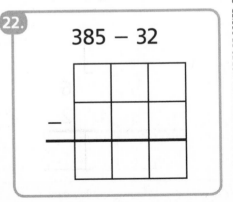

**20.** 627 − 410
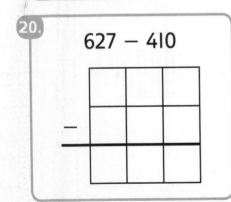

**21.** 593 − 302

**22.** 385 − 32

☐ Cross out blocks to show the number you take away.

☐ Show the regrouping in the numbers.

☐ Subtract.

**1.**

|  | 2 | 13 |  |
|---|---|---|---|
|  | 3̶ | 3̶ | 5 |
| − | 1 | 6 | 3 |
|  | 1 | 7 | 2 |

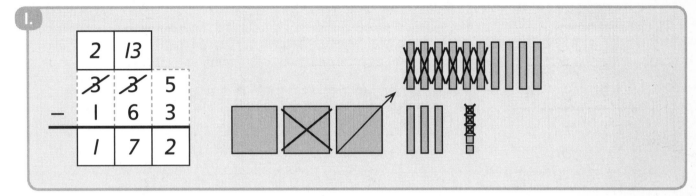

**2.**

|  |  |  |  |
|---|---|---|---|
|  | 4 | 5 | 6 |
| − | 1 | 8 | 2 |
|  |  |  |  |

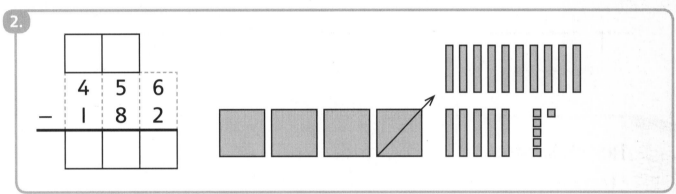

**3.**

|  |  |  |  |
|---|---|---|---|
|  | 5 | 2 | 7 |
| − | 2 | 5 | 2 |
|  |  |  |  |

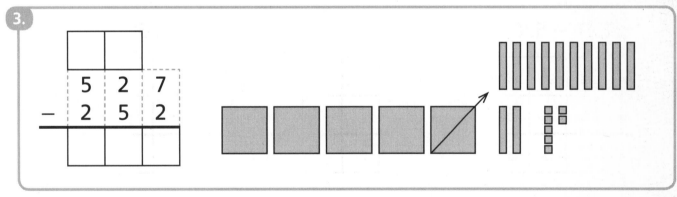

**4.**

|  |  |  |  |
|---|---|---|---|
|  | 4 | 1 | 6 |
| − | 1 | 4 | 2 |
|  |  |  |  |

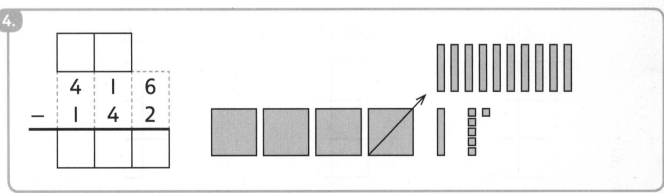

☐ Take apart a hundred.

☐ Show the regrouping in the numbers.

☐ Subtract.

**5.**

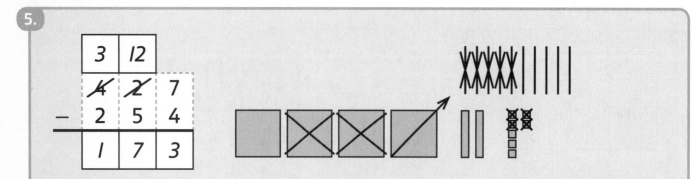

**6.**

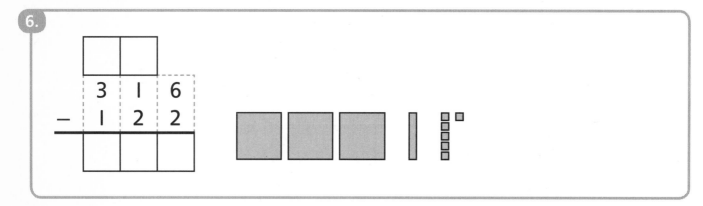

**7.**

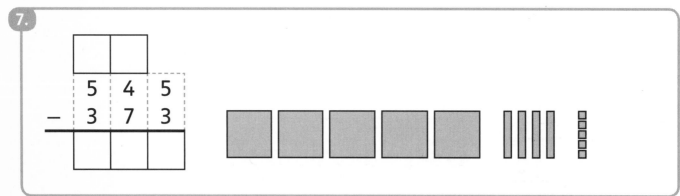

**8.**

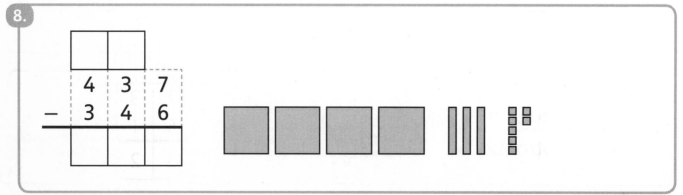

**Number and Operations in Base Ten 2-37**

☐ Draw a picture to show the subtraction.
☐ Subtract.

**9.**

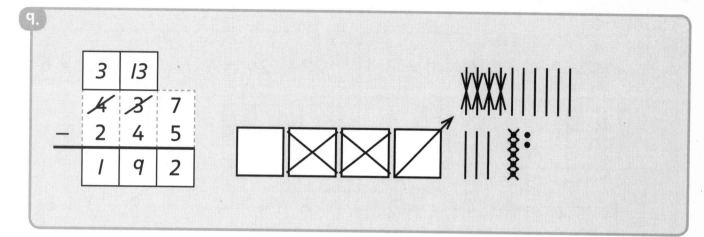

|   | 3 | 13 |   |
|---|---|----|---|
|   | 4̶ | 3̶  | 7 |
| − | 2 | 4  | 5 |
|   | 1 | 9  | 2 |

**10.**

|   |   |   |   |
|---|---|---|---|
|   | 5 | 3 | 4 |
| − | 2 | 7 | 1 |
|   |   |   |   |

**11.**

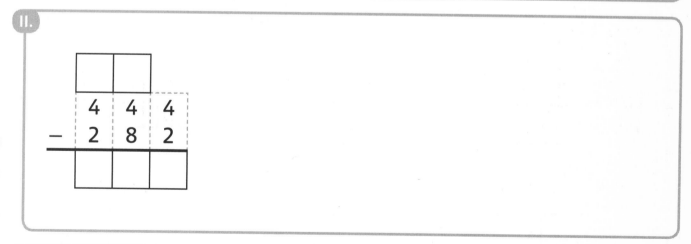

|   |   |   |   |
|---|---|---|---|
|   | 4 | 4 | 4 |
| − | 2 | 8 | 2 |
|   |   |   |   |

**12. BONUS**

What mistake did Amy make
in this subtraction?

|   | 7 | 2 | 3 |
|---|---|---|---|
| − | 2 | 5 | 1 |
|   | 5 | 3 | 2 |

The picture shows the first subtraction.

☐ Circle the part that shows the second subtraction.

**1.**

$17 - 15 = 2$

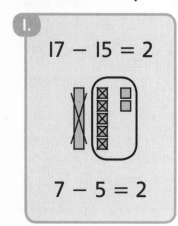

$7 - 5 = 2$

**2.**

$14 - 11 = 3$

$4 - 1 = 3$

**3.**

$28 - 24 = 4$

$8 - 4 = 4$

**4.**

$19 - 13 = 6$

$9 - 3 = 6$

The two subtractions have the same answer.

☐ Draw a picture to show why.

**5.**

$16 - 11 = 5$

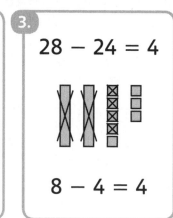

$6 - 1 = 5$

**6.**

$13 - 12 = 1$

$3 - 2 = 1$

**7.**

$25 - 23 = 2$

$5 - 3 = 2$

**8. BONUS**

$32 - 30 = 2$

$2 - 0 = 2$

☐ Draw a picture to show the subtraction.

☐ Circle the ones.

☐ Write the subtraction that the ones show.

**9.**

$28 - 23 = 5$

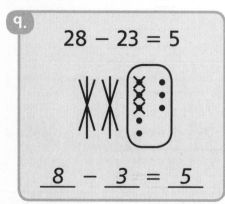

$\underline{8} - \underline{3} = \underline{5}$

**10.**

$14 - 12 = 2$

___ − ___ = ___

**11.**

$46 - 42 = 4$

___ − ___ = ___

The picture shows the first subtraction.

☐ Circle the tens and ones.

☐ Write the subtraction that the tens and ones show.

**12.**

237 − 214 = 23

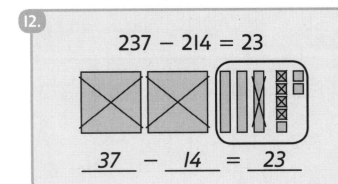

__37__ − __14__ = __23__

**13.**

156 − 125 = 31

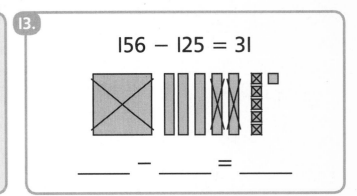

_____ − _____ = _____

**14.**

325 − 311 = 14

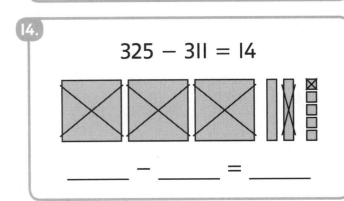

_____ − _____ = _____

**15.**

264 − 232 = 32

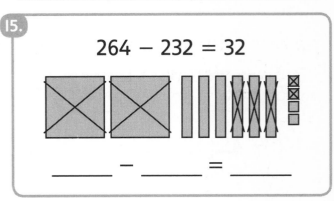

_____ − _____ = _____

☐ Write an easier subtraction.

☐ Subtract.

**16.**

17 − 15 = __7__ − __5__ = __2__

**17.**

39 − 36 = ___ − ___ = ___

**18.**

315 − 309 = __15__ − __9__ = ___

**19.**

761 − 758 = ___ − ___ = ___

**20.**

964 − 961 = ___ − ___ = ___

**21.**

422 − 418 = ___ − ___ = ___

**22. BONUS**

487 − 483 = _____ − _____ = _____ − _____ = _____

# NBT2-39 Using Addition to Subtract

◯ Fill in the missing numbers in the picture.

◯ Find the distances.

◯ Subtract by adding the distances.

**I.**

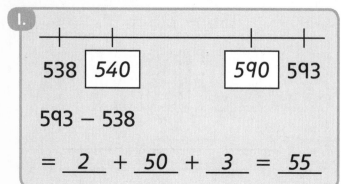

538 | 540 | | 590 | 593

593 − 538

= __2__ + __50__ + __3__ = __55__

**2.**

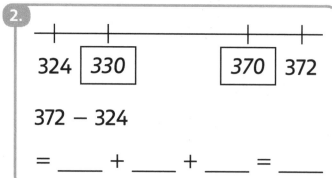

324 | 330 | | 370 | 372

372 − 324

= ___ + ___ + ___ = ___

**3.**

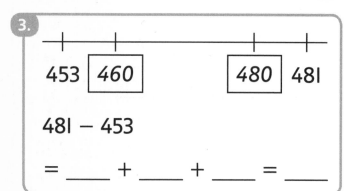

453 | 460 | | 480 | 481

481 − 453

= ___ + ___ + ___ = ___

**4.**

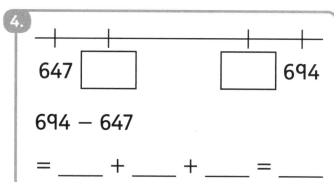

647 | | | | 694

694 − 647

= ___ + ___ + ___ = ___

**5.**

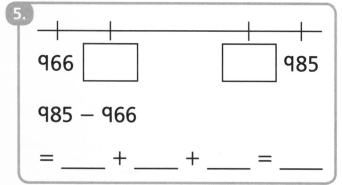

966 | | | | 985

985 − 966

= ___ + ___ + ___ = ___

**6.**

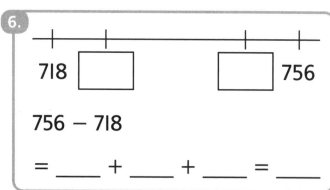

718 | | | | 756

756 − 718

= ___ + ___ + ___ = ___

**7. BONUS**

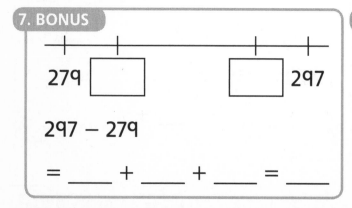

279 | | | | 297

297 − 279

= ___ + ___ + ___ = ___

**8. BONUS**

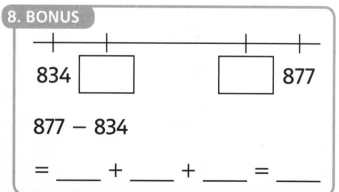

834 | | | | 877

877 − 834

= ___ + ___ + ___ = ___

☐ Find the distances.

☐ Subtract by adding the distances.

**9.**

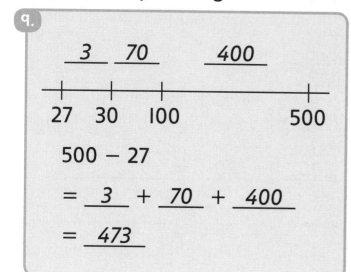

$$\underline{\quad 3 \quad} \quad \underline{\quad 70 \quad} \qquad \underline{\quad 400 \quad}$$

27　30　　100　　　　　500

500 − 27

= __3__ + __70__ + __400__

= __473__

**10.**

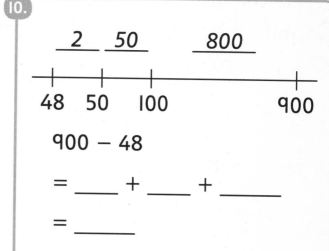

$$\underline{\quad 2 \quad} \quad \underline{\quad 50 \quad} \qquad \underline{\quad 800 \quad}$$

48　50　　100　　　　　900

900 − 48

= ___ + ___ + _____

= _____

**11.**

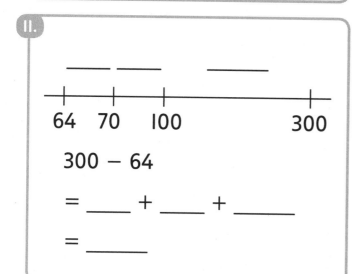

_____ _____ 　_____

64　70　　100　　　　　300

300 − 64

= ___ + ___ + _____

= _____

**12.**

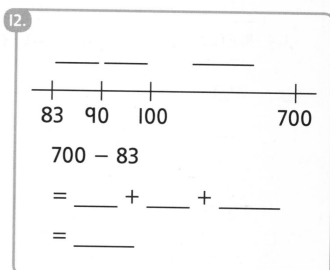

_____ _____ 　_____

83　90　　100　　　　　700

700 − 83

= ___ + ___ + _____

= _____

**13.**

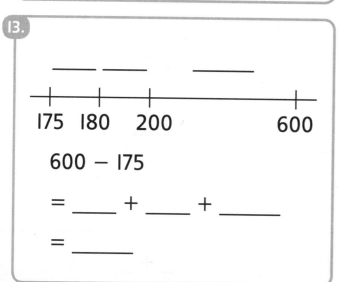

_____ _____ 　_____

175　180　　200　　　　　600

600 − 175

= ___ + ___ + _____

= _____

**14.**

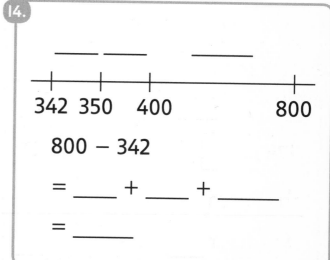

_____ _____ 　_____

342　350　　400　　　　　800

800 − 342

= ___ + ___ + _____

= _____

☐ Find the distances.

☐ Subtract by adding the distances.

**15.**

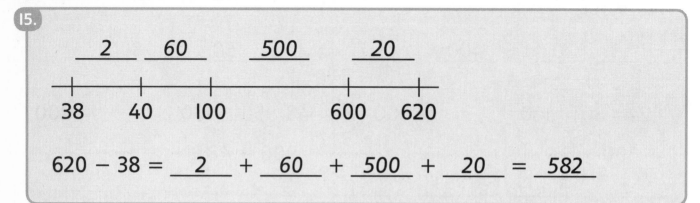

$$620 - 38 = \underline{\;2\;} + \underline{\;60\;} + \underline{\;500\;} + \underline{\;20\;} = \underline{\;582\;}$$

**16.**

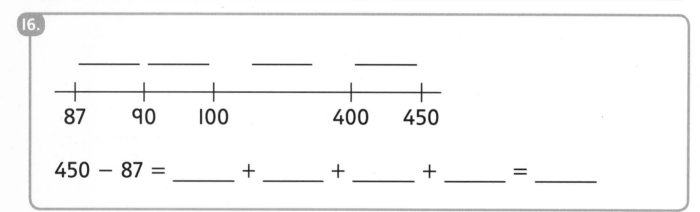

$$450 - 87 = \underline{\phantom{XX}} + \underline{\phantom{XX}} + \underline{\phantom{XX}} + \underline{\phantom{XX}} = \underline{\phantom{XX}}$$

**17.**

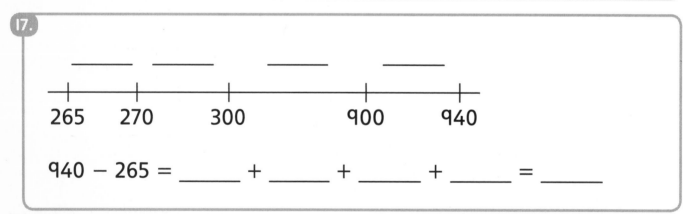

$$940 - 265 = \underline{\phantom{XX}} + \underline{\phantom{XX}} + \underline{\phantom{XX}} + \underline{\phantom{XX}} = \underline{\phantom{XX}}$$

**18. BONUS**

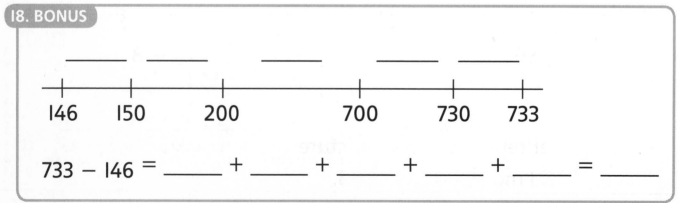

$$733 - 146 = \underline{\phantom{XX}} + \underline{\phantom{XX}} + \underline{\phantom{XX}} + \underline{\phantom{XX}} + \underline{\phantom{XX}} = \underline{\phantom{XX}}$$

Number and Operations in Base Ten 2-39

☐ Circle the subtraction that does **not** need regrouping.

☐ Subtract.

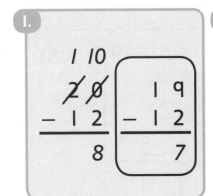

**1.**

$$\begin{array}{r} \overset{\scriptstyle 1\;10}{\cancel{2\,0}} \\ -\ 1\ 2 \\ \hline 8 \end{array} \qquad \begin{array}{r} 1\ 9 \\ -\ 1\ 2 \\ \hline 7 \end{array}$$

**2.**

$$\begin{array}{r} 6\ 9 \\ -\ 3\ 7 \\ \hline 3\ 2 \end{array} \qquad \begin{array}{r} \overset{\scriptstyle 6\;10}{\cancel{7\,0}} \\ -\ 3\ 7 \\ \hline 3\ 3 \end{array}$$

**3.**

$$\begin{array}{r} 4\ 0 \\ -\ 1\ 4 \\ \hline \end{array} \qquad \begin{array}{r} 3\ 9 \\ -\ 1\ 4 \\ \hline \end{array}$$

☐ Subtract to find the distances in the picture.

☐ Subtract by adding the distances.

**4.**

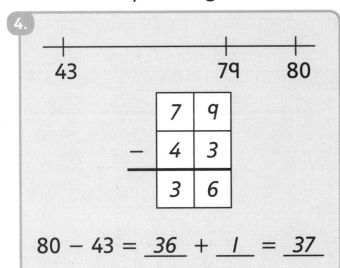

43         79   80

$$\begin{array}{r} 7\ 9 \\ -\ 4\ 3 \\ \hline 3\ 6 \end{array}$$

$80 - 43 = \underline{\ 36\ } + \underline{\ 1\ } = \underline{\ 37\ }$

**5.**

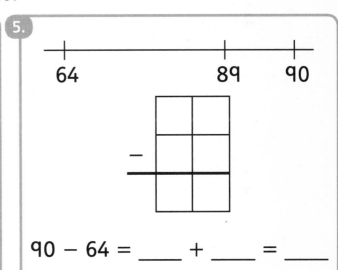

64         89   90

$90 - 64 = \underline{\ \ \ } + \underline{\ \ \ } = \underline{\ \ \ }$

**6.**

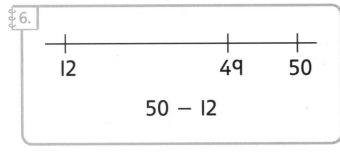

12         49   50

$50 - 12$

**7.**

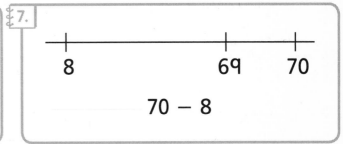

8         69   70

$70 - 8$

**8. BONUS**

Write a sentence or draw a picture to show why

$70 - 33$ is 1 more than $69 - 33$.

Subtract to find the distances in the picture.
Subtract by adding the distances.

**9.**

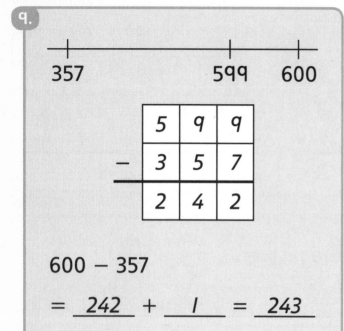

357        599    600

$$
\begin{array}{r}
5\ 9\ 9 \\
-\ 3\ 5\ 7 \\
\hline
2\ 4\ 2
\end{array}
$$

600 − 357

= __242__ + __1__ = __243__

**10.**

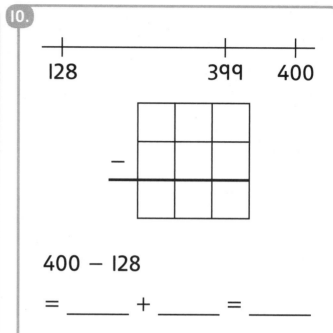

128        399    400

400 − 128

= _____ + _____ = _____

**11.**

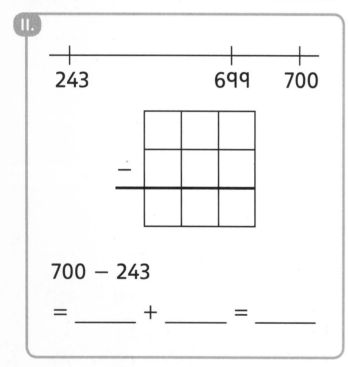

243        699    700

700 − 243

= _____ + _____ = _____

**12.**

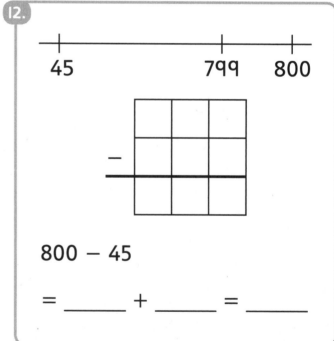

45        799    800

800 − 45

= _____ + _____ = _____

Subtract. Use any way you like.

**13.**

300 − 127

**14.**

600 − 346

# OA2-49  Counting by 2s

☐ Count by 2s to find the number of circles.

**1.**

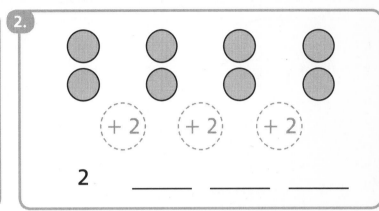

2      4

**2.**

2 ____ ____ ____

**3.**

2 ____ ____

**4.**

2 ____ ____ ____ ____

**5.**

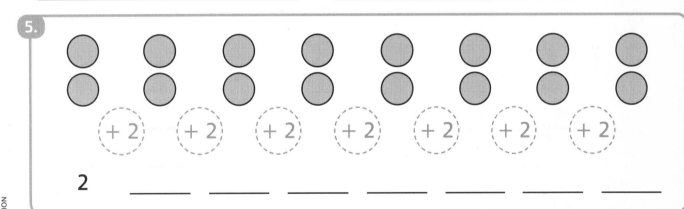

2 ____ ____ ____ ____ ____ ____ ____

**6.**

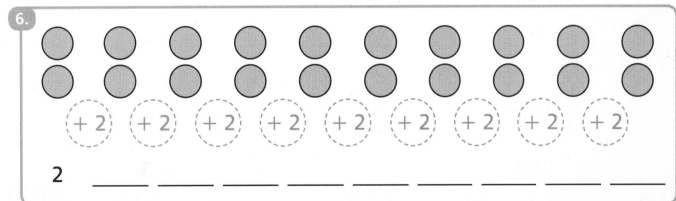

2 ____ ____ ____ ____ ____ ____ ____ ____ ____

☐ Count the circles in each row to write an addition.

**7.**
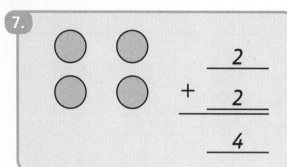

$$
\begin{array}{r}
2 \\
+ \quad 2 \\
\hline
4
\end{array}
$$

**8.**
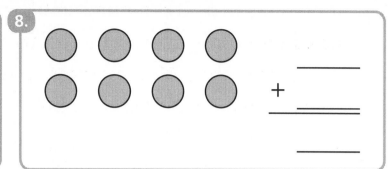

$+$ _____

**9.**
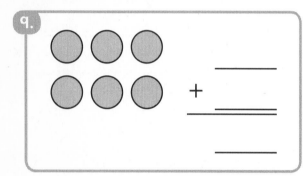

$+$ _____

**10.**
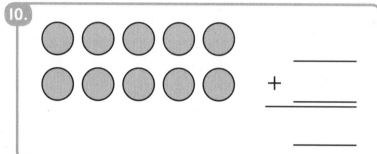

$+$ _____

**11.**
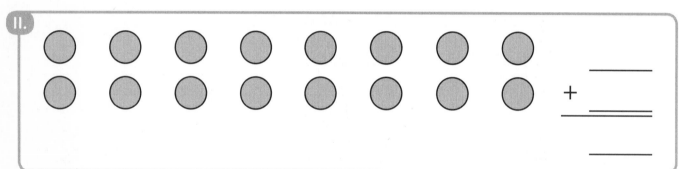

$+$ _____

**12.**
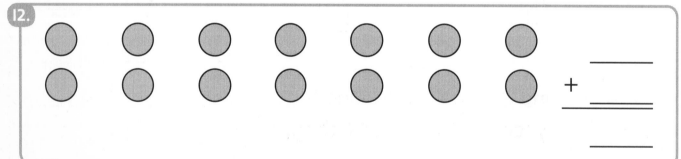

$+$ _____

**13.**

$+$ _____

◯ Write an addition to find the total.

**14.**

Alexa brought 6 green marbles to school.
Marco brought 6 blue marbles to school.
How many marbles did they bring altogether?

$$\begin{array}{r} 6 \\ + \phantom{0}6 \\ \hline 12 \end{array}$$

**15.**

Raj bought 7 cartons of orange juice and 7 cartons of milk.
How many cartons did he buy altogether?

$$\begin{array}{r} \underline{\phantom{00}} \\ + \phantom{0} \\ \hline \phantom{00} \end{array}$$

**16.**

Tasha found 8 aspen leaves and 8 oak leaves.
How many leaves did she find altogether?

$$\begin{array}{r} \underline{\phantom{00}} \\ + \phantom{0} \\ \hline \phantom{00} \end{array}$$

☐ Count by 3s to find the total.

**1.**

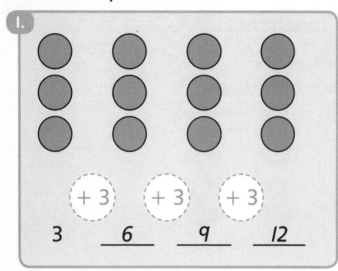

3    __6__    __9__    __12__

**2.**

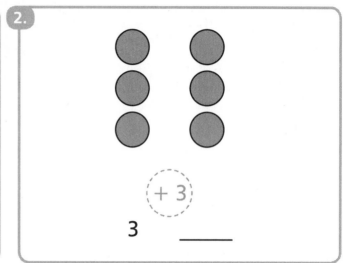

3    _____

**3.**

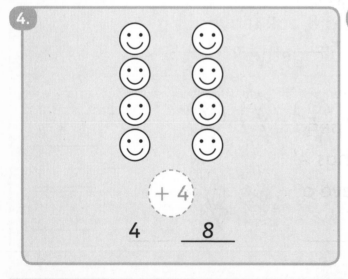

3    _____    _____    _____    _____

☐ Count by 4s to find the total.

**4.**

4    __8__

**5.**

4    _____    _____

◯ Count by 5s to find the total.

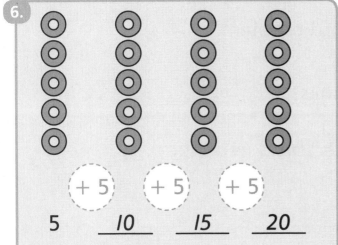

6.

5    _10_    _15_    _20_

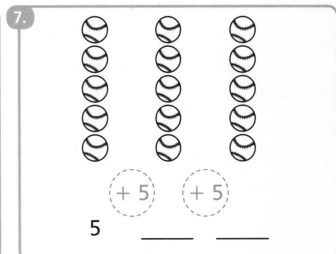

7.

5    ____    ____

◯ Write an equation. Then find the total.

8.

Rob counts 3 green cars, 3 blue cars, 3 black cars,
3 yellow cars, and 3 red cars. How many cars does
he count altogether?

_____ + _____ + _____ + _____ + _____ = _____

9.

Randi has 5 green marbles and 5 blue marbles.
Her friend gives her 5 red marbles.
How many marbles does she have altogether?

_____ + _____ + _____ = _____

10.

Sam has 4 pennies, Tess has 4 pennies,
Nina has 4 pennies, and Jake has 4 pennies.
How many pennies do they have altogether?

_____ + _____ + _____ + _____ = _____

# OA2-51 Two-Step Word Problems with Result Unknown

☐ Underline the numbers in the word problem.

☐ Write + or − in each circle.

☐ Use the numbers to fill in the blanks.

**1.**

There were <u>9</u> blue balls and <u>5</u> red balls in the bag.
Cathy put in <u>7</u> more balls.

<u>  9  </u> ( + ) <u>  5  </u> ( + ) <u>  7  </u>

**2.**

Emma had 3 short pencils and 6 long pencils.
The teacher gave her 8 more pencils.

___ ◯ ___ ◯ ___

**3.**

Ray swam across the pool 17 times on his front and 8 times
on his back. Then he swam across 4 more times on his side.

___ ◯ ___ ◯ ___

**4. BONUS**

Zack chopped 59 carrots and 83 onions. Then he
chopped 104 more onions.

___ ◯ ___ ◯ ___

◯ Underline the numbers in the word problem.

◯ Write + or − in each circle.

◯ Use the numbers to fill in the blanks.

**5.**

There were <u>9</u> plums on the plate. The friends ate <u>4</u> plums.
Later they ate <u>3</u> more plums.

__9__ $\left(-\right)$ __4__ $\left(-\right)$ __3__

**6.**

There were 15 birds on the gate. 5 birds flew away.
Then 2 more birds flew away.

_____ ◯ _____ ◯ _____

**7.**

There were 10 beavers in the pond. 2 beavers swam away.
Then 5 more beavers swam away.

_____ ◯ _____ ◯ _____

**8. BONUS**

There were 382 flies near the pond. 153 flies flew away.
Then 68 more flies flew away.

_____ ◯ _____ ◯ _____

☐ Underline the numbers in the word problem.
☐ Write + or − in each circle.
☐ Use the numbers to fill in the blanks. Then find the answer.

**9.**

There were <u>16</u> robins on the roof. <u>5</u> robins flew away.
Then <u>8</u> more robins landed on the roof. How many
robins are on the roof now?

$$\underline{\ 16\ } \ \bigcirc{-} \ \underline{\ 5\ } \ \bigcirc{+} \ \underline{\ 8\ } \ = \ \underline{\ 19\ }$$

**10.**

Sara got 7 books from the library. Then she got
6 more books and took 2 books back. How many
books does she have now?

$$\underline{\ \ \ \ } \ \bigcirc \ \underline{\ \ \ \ } \ \bigcirc \ \underline{\ \ \ \ } \ = \ \underline{\ \ \ \ }$$

**11.**

Jay made 14 cookies. Then he made 12 more cookies
and gave 21 cookies away. How many cookies does
he have now?

$$\underline{\ \ \ \ } \ \bigcirc \ \underline{\ \ \ \ } \ \bigcirc \ \underline{\ \ \ \ } \ = \ \underline{\ \ \ \ }$$

**12. BONUS**

A teacher bought 37 markers for his class. He gave
29 markers to his students and bought 58 more markers.
How many markers does he have now?

$$\underline{\ \ \ \ } \ \bigcirc \ \underline{\ \ \ \ } \ \bigcirc \ \underline{\ \ \ \ } \ = \ \underline{\ \ \ \ }$$

Operations and Algebraic Thinking 2-51

☐ Fill in the blanks using the numbers in the word problem.
☐ Write + or − in each circle.
☐ Find the answer.

**13.**
There were 13 baby turtles in the nest. 3 turtles crawled away and then 2 more crawled away. How many turtles are there now?

__13__ ⊖ __3__ ⊖ __2__ = __8__

**14.**
There were 10 pears in the bag. Amit put 7 more pears in the bag and then he took out 2. How many pears are in the bag now?

_____ ◯ _____ ◯ _____ = _____

**15.**
There were 22 cards in the pile. Jen took away 2 cards and Bo put 4 more cards on the pile. How many cards are in the pile now?

_____ ◯ _____ ◯ _____ = _____

**16.**
There were 81 coins in the jar. Peter put 9 more coins in the jar and Milly took out 40 coins. How many coins are in the jar now?

**17.**
There were 45 tacks in the box. Fred took out 5 tacks. Liz put 17 more tacks in the box. How many tacks are in the box now?

# OA2-52 Number-Bond and Part-Whole Pictures

☐ Count the circles in each group.
☐ Write the numbers.

**1.**
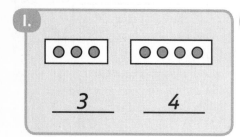

_____3_____     _____4_____

**2.**
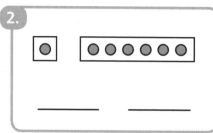

_____     _____

**3.**
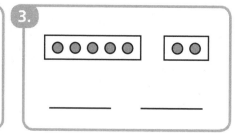

_____     _____

☐ Write the total number of circles.

**4.**
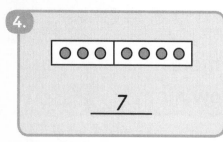

_____7_____

**5.**
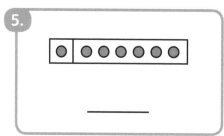

_____

**6.**
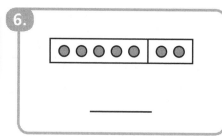

_____

☐ Write the total. Then write the number in each group.

**7.**
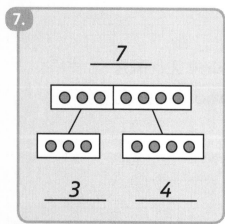

_____3_____     _____4_____

**8.**
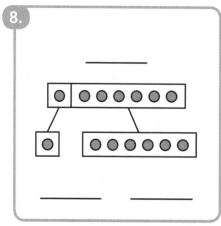

_____     _____

**9.**
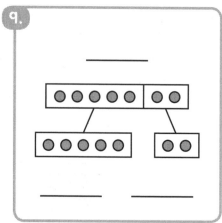

_____     _____

**10. BONUS**

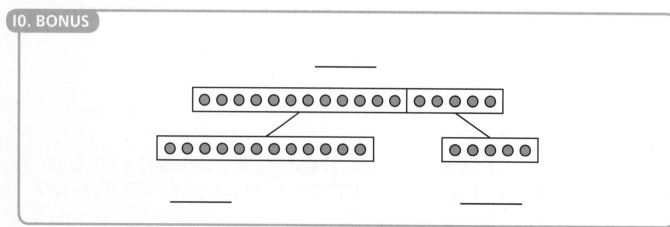

◯ Draw the missing circles.

11.

12.

13.

14.

15.

16.

17.

Operations and Algebraic Thinking 2-52

◯ Write the addition. Draw ☐ for the number you do not know.

**18.**

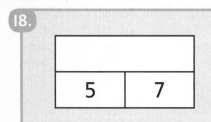

|  |  |
|---|---|
| 5 | 7 |

5 + 7 = ☐

**19.**

|  14 |  |
|---|---|
| 9 |  |

9 + ☐ = 14

**20.**

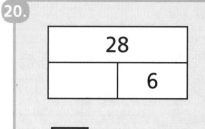

|  28 |  |
|---|---|
|  | 6 |

☐ + 6 = 28

**21.**

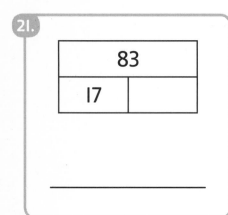

|  83 |  |
|---|---|
| 17 |  |

_____

**22.**

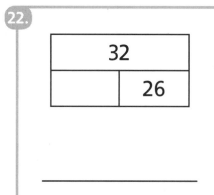

|  32 |  |
|---|---|
|  | 26 |

_____

**23.**

|  |  |
|---|---|
| 19 | 25 |

_____

**24.**

|  49 |  |
|---|---|
|  | 22 |

_____

**25.**

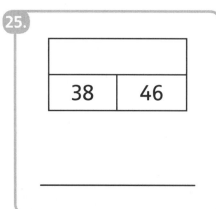

|  |  |
|---|---|
| 38 | 46 |

_____

**26.**

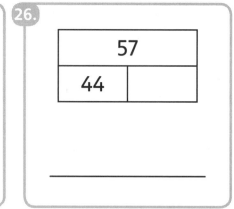

|  57 |  |
|---|---|
| 44 |  |

_____

**27.**

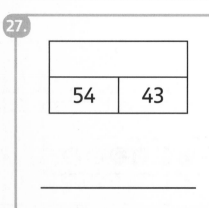

|  |  |
|---|---|
| 54 | 43 |

_____

**28.**

|  81 |  |
|---|---|
| 63 |  |

_____

**29.**

|  95 |  |
|---|---|
|  | 72 |

_____

○ Add or subtract to find the missing number.

**30.**

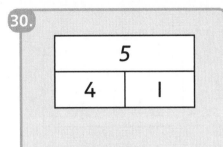

| 5 | |
|---|---|
| 4 | 1 |

___4 + 1 = 5___

**31.**

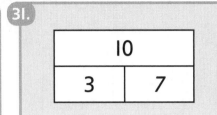

| 10 | |
|---|---|
| 3 | 7 |

___10 − 3 = 7___

**32.**

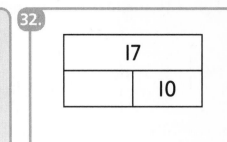

| 17 | |
|---|---|
| | 10 |

_____

**33.**

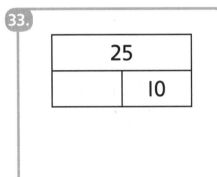

| 25 | |
|---|---|
| | 10 |

_____

**34.**

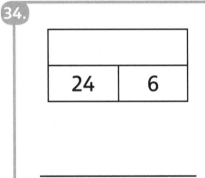

| | |
|---|---|
| 24 | 6 |

_____

**35.**

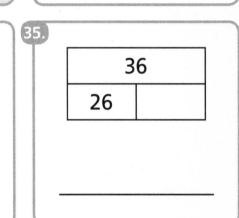

| 36 | |
|---|---|
| 26 | |

_____

**36.**

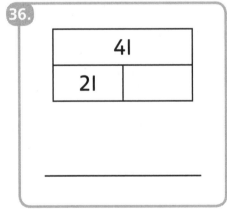

| 41 | |
|---|---|
| 21 | |

_____

**37.**

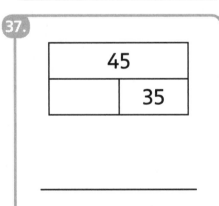

| 45 | |
|---|---|
| | 35 |

_____

**38.**

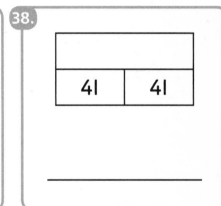

| | |
|---|---|
| 41 | 41 |

_____

**39.**

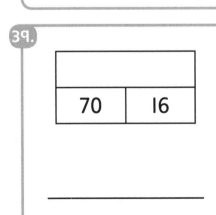

| | |
|---|---|
| 70 | 16 |

_____

**40.**

| 90 | |
|---|---|
| 88 | |

_____

**41.**

| 68 | |
|---|---|
| | 58 |

_____

**Operations and Algebraic Thinking 2-52**

# OA2-53 Using Part-Whole Pictures to Write Equations

☐ Use the picture to write two subtraction sentences.

**1.**

| 18 | |
|---|---|
| 5 | 13 |

<u>18</u> – <u>5</u> = <u>13</u>

<u>18</u> – <u>13</u> = <u>5</u>

**2.**

| 63 | |
|---|---|
| 14 | 49 |

____ – ____ = ____

____ – ____ = ____

**3.**

| 77 | |
|---|---|
| 39 | 38 |

____ – ____ = ____

____ – ____ = ____

☐ Use the picture to write two subtractions.

**4.**

| 56 | |
|---|---|
| 31 | 25 |

|   | 5 | 6 |
|---|---|---|
| – | 3 | 1 |
|   | 2 | 5 |

|   | 5 | 6 |
|---|---|---|
| – | 2 | 5 |
|   | 3 | 1 |

**5.**

| 80 | |
|---|---|
| 47 | 33 |

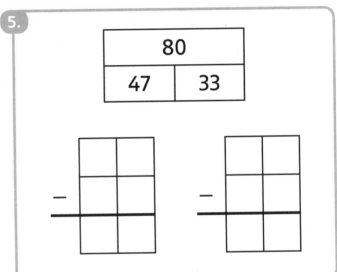

**6.**

| 76 | |
|---|---|
| 25 | 51 |

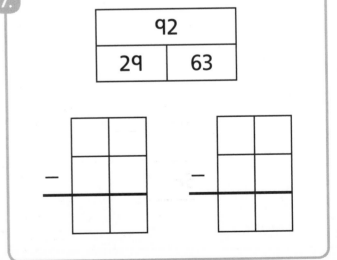

**7.**

| 92 | |
|---|---|
| 29 | 63 |

COPYRIGHT © 2014 JUMP MATH: NOT TO BE COPIED. CC EDITION

☐ Use the grid to write the numbers you know.

**8.**

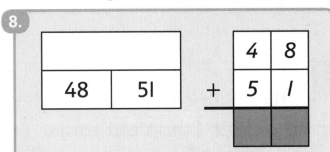

|    |    |
|----|----|
|    |    |
| 48 | 51 |

|   | 4 | 8 |
|---|---|---|
| + | 5 | 1 |
|   |   |   |

**9.**

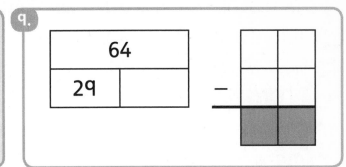

|    |  |
|----|--|
| 64 |  |
| 29 |  |

| − |   |   |
|---|---|---|
|   |   |   |

**10.**

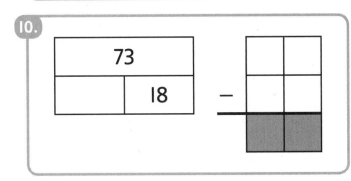

|    |    |
|----|----|
| 73 |    |
|    | 18 |

| − |   |   |
|---|---|---|
|   |   |   |

**11.**

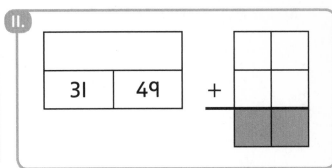

|    |    |
|----|----|
|    |    |
| 31 | 49 |

| + |   |   |
|---|---|---|
|   |   |   |

☐ Use the grid to write the numbers you know.

☐ Write + or −. Then add or subtract.

☐ Fill in the picture.

**12.**

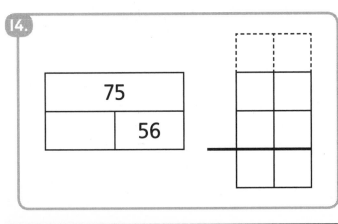

|    |    |
|----|----|
| 82 |    |
| 67 | 15 |

|   | 7 | 12 |
|---|---|----|
|   | 8̸ | 2̸  |
| − | 6 | 7  |
|   | 1 | 5  |

**13.**

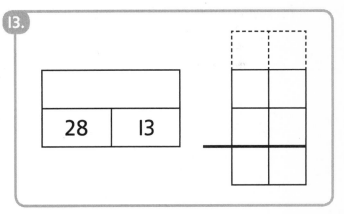

|    |    |
|----|----|
|    |    |
| 28 | 13 |

**14.**

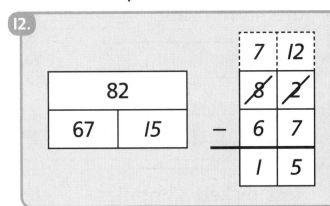

|    |    |
|----|----|
| 75 |    |
|    | 56 |

**15.**

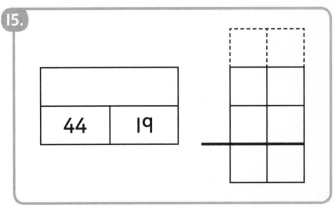

|    |    |
|----|----|
|    |    |
| 44 | 19 |

# OA2-54 One-Step Word Problems with Change Unknown

☐ Fill in the numbers you know.

**1.** There were 14 nuts in the bowl. Clara put some more nuts in the bowl. Now there are 29 nuts.

| 29 | |
|----|----|
| 14 | |

**2.** There were 32 cups on the shelf. Ethan put some more cups on the shelf. Now there are 51 cups.

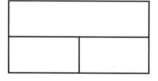

**3.** There were 25 children at the beach. Some more children went to the beach. Now there are 61 children.

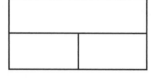

**4.** Mandy made 36 muffins. Then she made some more muffins. Now there are 48 muffins.

**5.** Andy typed 17 names. Then he typed some more names. Now there are 55 names.

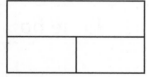

**6.** Tess picked 38 apples. Then she picked some more apples. Now she has 47 apples.

○ Fill in the numbers you know.
○ Subtract.
○ Write the answer in the blanks.

**7.**

There were 23 cows in the field. Some more cows came. Now there are 32 cows. How many more cows came to the field?

| 32 | |
|---|---|
| 23 | 9 |

$$
\begin{array}{c|c}
2 & 12 \\
\hline
\cancel{3} & \cancel{2} \\
\hline
- \quad 2 & 3 \\
\hline
 & 9 \\
\end{array}
$$

___9___ more cows came
to the field.

**8.**

There were 18 people in the pool. Some more came. Now there are 31 people. How many more people came to the pool?

| | |
|---|---|
| | |

_____ more people
came to the pool.

**9.**

There were 39 blocks in the bag. Jack put some more in the bag. Now there are 55 blocks. How many more blocks did Jack put in?

| | |
|---|---|
| | |

Jack put _____ more
blocks in the bag.

subtract using the numbers you know.

○ Write the answer in a word sentence.

**10.** There were 29 kites in the store. The owner brought some more kites. Now there are 47 kites. How many kites did she bring?

**11.** Abdul had 43 sea shells. He found some more sea shells. Now he has 62 sea shells. How many sea shells did he find?

**12.** Grace knew the names of 31 Olympic athletes. Then she learned some more names. Now she knows 50 names of Olympic athletes. How many more names did she learn?

**13.** Lynn walked by 47 stores. Then she walked by some more stores. Altogether she walked by 63 stores. How many more stores did she walk by?

**14.** Sal learned 35 riddles. Then he learned some more riddles. Now he knows 49 riddles. How many more riddles did he learn?

**15.** Alice counted 88 seconds on the clock. Then she counted some more seconds. Altogether she counted 95 seconds. How many more seconds did she count?

Operations and Algebraic Thinking 2-54

☐ Fill in the numbers you know.
☐ Subtract.

**1.**

There were 32 lights on.
Blanca turned off some lights.
Now 3 lights are on. How many
lights did Blanca turn off?

| | 32 |
|---|---|
| | 3 |

$$\begin{array}{c c} \overset{2}{\cancel{3}} & \overset{12}{\cancel{2}} \\ - \quad & 3 \\ \hline 2 & 9 \end{array}$$

**2.**

There are 46 straws in the box.
19 straws are blue and the rest
are green. How many straws
are green?

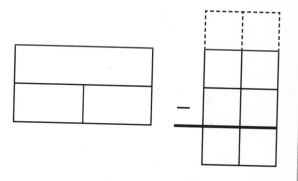

**3.**

There were 25 open books.
Ed closed some of the books.
Now there are 17 open books.
How many books did he close?

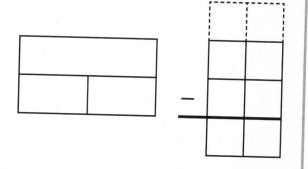

ill in the numbers you know. Then subtract.

**4.** There are 67 starfish. 48 starfish are red and the rest are pink. How many starfish are pink?

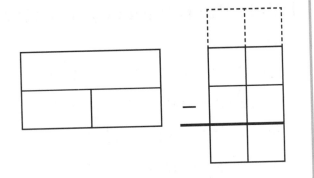

**5.** There were 33 children sitting. Some children stood up. Now there are 25 children sitting. How many children stood up?

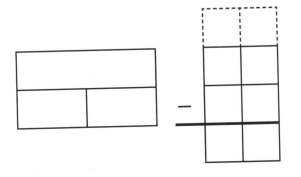

**6.** There are 43 butterflies. 28 butterflies are orange and the rest are yellow. How many butterflies are yellow?

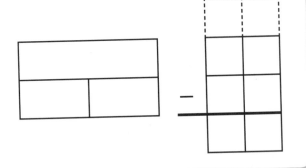

**7.** There are 73 balloons. 29 balloons are big and the rest are small. How many balloons are small?

**8.** There are 52 boxes on the floor. Some boxes are closed and 15 are open. How many boxes are closed?

Where do the numbers go in the picture?
☐ Write the numbers you know.

**1.**
Anne has 13 red stickers and 12 blue stickers. How many stickers does Anne have altogether?

**2.**
Greg drew 19 circles and some squares. He drew 25 shapes altogether. How many squares did he draw?

**3.**
Ivan has 18 grapes. 7 are green and the rest are purple. How many grapes are purple?

**4.**
Sun wrote 37 words at home. She wrote 42 words at school. How many words did she write altogether?

**5.**
An owl laid 8 eggs last year. She laid 11 eggs this year. How many eggs did the owl lay altogether?

**6.**
The teacher has 9 letters in her name. 4 letters are vowels. How many letters are not vowels?

Use the picture to write the numbers you know.
Write + or − in the circle. Then add or subtract.

**7.**

Hanna used 3 stamps last week and 7 stamps this week. How many stamps did she use altogether?

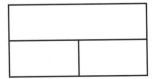

\_\_\_\_\_ ◯ \_\_\_\_\_ = \_\_\_\_\_

**8.**

There were 10 fish under a rock. 6 fish swam away. How many fish are under the rock now?

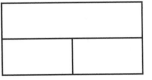

\_\_\_\_\_ ◯ \_\_\_\_\_ = \_\_\_\_\_

**9.**

A squirrel found 12 peanuts and some pine nuts. The squirrel found 15 nuts altogether. How many nuts were pine nuts?

\_\_\_\_\_ ◯ \_\_\_\_\_ = \_\_\_\_\_

**10.**

A mother gorilla ate 9 bananas. Her baby ate 3 bananas. How many bananas did they eat altogether?

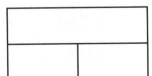

\_\_\_\_\_ ◯ \_\_\_\_\_ = \_\_\_\_\_

# OA2-57 Two-Step Word Problems with Missing Addends

○ Subtract using the numbers you know.
○ Use the answer from the subtraction to complete the question.

**1.**

There were 8 blue stars and some red stars. There were 12 stars altogether. Then the students made 6 more red stars. How many red stars are there now?

| 12 | |
|----|----|
| 8 | 4 |

__12__ − __8__ = [ 4 ]

__4__ + __6__ = __10__

There are __10__ red stars now.

**2.**

There were 13 tree frogs and some bullfrogs at the zoo. There were 20 frogs altogether. Then the zoo got 5 more bullfrogs. How many bullfrogs are there now?

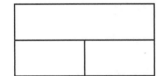

_____ − _____ = [  ]

_____ + _____ = _____

There are_____ bullfrogs now.

**3.**

Tim had 16 old cards and some new cards. He had 26 cards altogether. Then Tim got 4 more new cards. How many new cards does he have now?

_____ − _____ = [  ]

_____ + _____ = _____

Tim has _____ new cards now.

☐ Subtract using the numbers you know.
☐ Use the answer from the subtraction to complete the question.

**4.**

Mary has 20 small sheep and some big sheep. She has 26 sheep altogether. Then she bought 7 more big sheep. How many big sheep does Mary have in total?

| 26 | |
|----|----|
| 20 | 6 |

$\underline{\quad 26 - 20 \quad} = \boxed{6}$

$\underline{\quad 6 + 7 = 13 \quad}$

Mary has __13__ big sheep.

**5.**

Kyle ran 25 short races and some long races. He ran 35 races altogether. Then Kyle ran 9 more long races. How many long races did he run in total?

| | |
|----|----|
| | |

_____

_____

Kyle ran _____ long races.

**6.**

Yu collected 18 gray rocks and some brown rocks. She collected 29 rocks altogether. Then Yu collected 2 more brown rocks. How many brown rocks did she collect in total?

**7.**

Jake trained 33 puppies and some adult dogs. He trained 40 dogs altogether. Then he trained 7 more adult dogs. How many adult dogs did Jake train in total?

**One-Step Word Problems with Comparing to Find the Difference**

☐ In the top row, draw circles for the larger number.
☐ In the bottom row, draw circles for the smaller number.
☐ Draw Xs in the empty boxes until the bottom equals the top.
☐ Write the number of Xs.

**1.**

Jane has 2. Mark has 6.

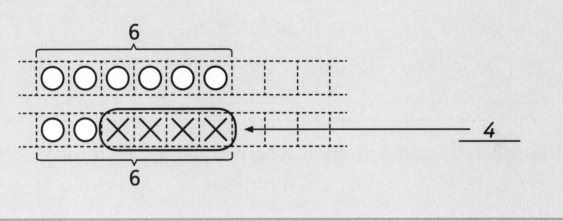

4

**2.**

Jim has 7. Amy has 3.

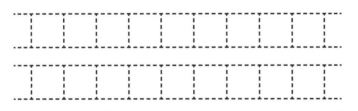

_____

**3.**

Beth has 2. Glen has 8.

_____

☐ Draw circles to show the numbers.

☐ Use the number of ✗s to complete the sentences.

**4.**

Jennifer picked 5 oranges. Will picked 8 oranges.

◯◯◯◯◯◯◯

◯◯◯◯◯✗✗✗

Will picked __3__ more than Jennifer.

Jennifer picked __3__ fewer than Will.

**5.**

Helen collected 4 stamps. Karen collected 6 stamps.

Karen collected _____ more than Helen.

Helen collected _____ fewer than Karen.

**6.**

David counted 7 cups. Tony counted 5 cups.

David counted _____ more than Tony.

Tony counted _____ fewer than David.

○ Write the larger number in the top box.
○ Write the smaller number in the bottom box.
What number makes the bottom equal the top?
○ Write the number in the oval.

**7.**

Mark caught 6 fish. Bev caught 8 fish.

| 8 | |
|---|---|
| 6 | ( 2 ) |

**8.**

Anwar wrote 2 e-mails. Nina wrote 5 e-mails.

| | |
|---|---|
| | ( ) |

**9.**

Vicky listened to 7 songs. Carlos listened to 3 songs.

| | |
|---|---|
| | ( ) |

**10.**

Scott read 4 stories. Lynn read 10 stories.

| | |
|---|---|
| | ( ) |

THE
THREE
BEARS

☐ Fill in the picture.

☐ Complete the sentences.

**11.**

Kate played 4 games at the fun fair.

Ben played 9 games.

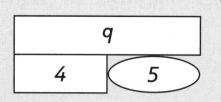

Ben played ___5___ more games than Kate.

Kate played ___5___ fewer games than Ben.

**12.**

Tom planted 10 seeds.

Sara planted 7 seeds.

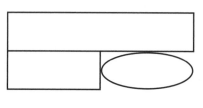

Tom planted _____ more seeds than Sara.

Sara planted _____ fewer seeds than Tom.

**13.**

Peter used 10 blocks to build a tower.

Jayden used 15 blocks.

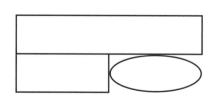

Jayden used _____ more blocks than Peter.

Peter used _____ fewer blocks than Jayden.

Operations and Algebraic Thinking 2-58

○ Circle who has more. Underline who has fewer.
○ Who has more? Write the name on the top line.
○ Who has fewer? Write the name on the bottom line.
○ Fill in the numbers you know.

**1.**

(Roy) has 3 more blocks than <u>Marta</u>. Marta has 8 blocks.

<u>   Roy   </u>

<u>  Marta  </u>

| | |
|---|---|
| 8 | (3) |

**2.**

Lily has 5 blocks. Mike has 6 more blocks than Lily.

_____

_____

**3.**

Ann has 17 blocks. Pat has 3 more blocks than Ann.

_____

_____

**4. BONUS**

Ravi has 76 more blocks than Nancy. Nancy has 21 blocks.

_____

_____

☐ Circle who has more. Underline who has fewer.

☐ Fill in what you know.

☐ Add to find the number you do not know.

☐ Complete the sentence.

**5.**

(Kim) has 4 more blocks than <u>John</u>. John has 2 blocks.

How many blocks does Kim have?

| Kim | | 6 | |
|-----|---|---|---|
| John | | 2 | 4 |

Kim has __6__ blocks.

**6.**

Rick has 2 more blocks than Kathy. Kathy has 6 blocks.

How many blocks does Rick have?

Rick has _____ blocks.

**7.**

Sally has 3 more blocks than Bobby. Bobby has 5 blocks.

How many blocks does Sally have?

Sally has _____ blocks.

Operations and Algebraic Thinking 2-59

○ Circle who has more. Underline who has fewer.
○ Fill in what you know.

**8.**

(Carlos) has 9 stickers. <u>Tessa</u> has 4 fewer stickers than Carlos.

Carlos _____

Tessa _____

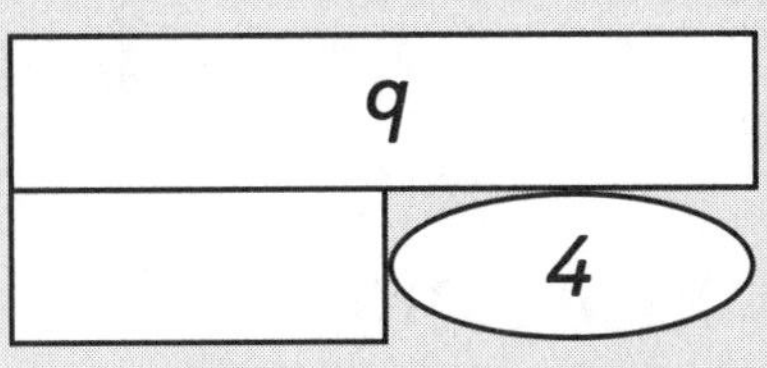

| | 9 |
|---|---|
| | (4) |

**9.**

Alex has 6 fewer stickers than Rob. Rob has 12 stickers.

_____
_____

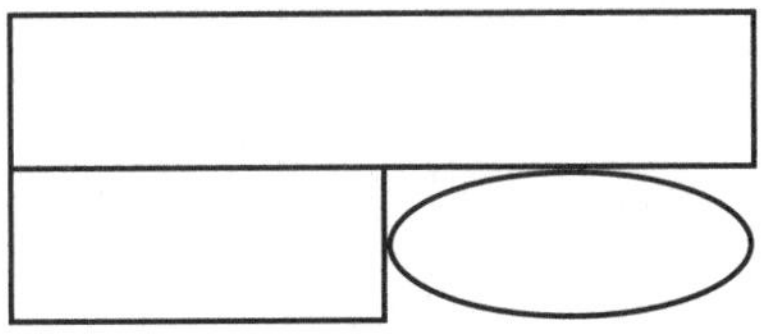

**10.**

Sam has 11 stickers. Marla has 9 fewer stickers than Sam.

_____
_____

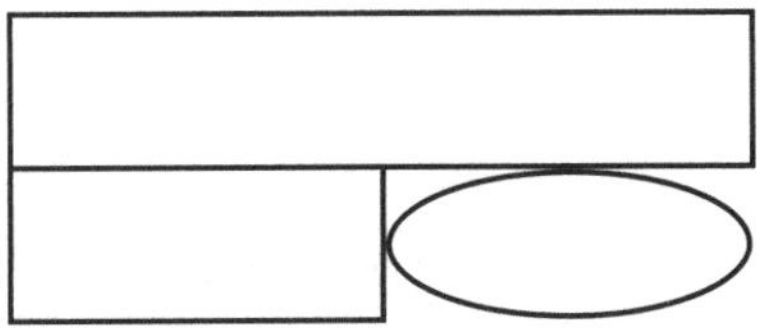

**11. BONUS**

Randi has 62 fewer stickers than Ray. Ray has 83 stickers.

_____
_____

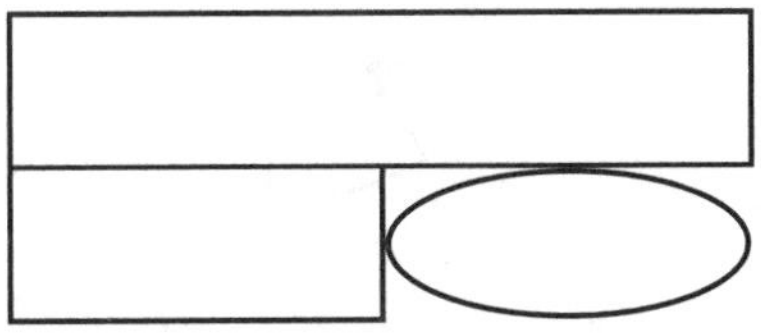

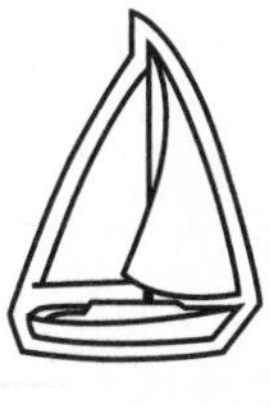

☐ Fill in what you know.
☐ Subtract to find the number you do not know.
☐ Complete the sentence.

**12.**

Sun has 4 fewer stickers than (Cam).

Cam has 9 stickers.

How many stickers does Sun have?

| Cam | 9 | |
| Sun | 5 | (4) |

Sun has ___5___ stickers.

**13.**

Hanna has 5 fewer stickers than Raj.

Raj has 12 stickers.

How many stickers does Hanna have?

Hanna has _____ stickers.

**14.**

Pam has 3 fewer stickers than Bob.

Bob has 10 stickers.

How many stickers does Pam have?

Pam has _____ stickers.

Operations and Algebraic Thinking 2-59

Write + or − in the circle. Then add or subtract.

**15.**

Clara has 27 fewer coins than (Jin.)

Jin has 44 coins.

How many coins does Clara have?

| 44 | |
|---|---|
| 17 | (27) |

|  | 3 | 14 |
|---|---|---|
| − | 4̷ | 4̷ |
|  | 2 | 7 |
|  | 1 | 7 |

**16.**

Ava has 19 more coins than Jenny.

Jenny has 38 coins.

How many coins does Ava have?

**17.**

Ross has 52 more coins than Don.

Don has 29 coins.

How many coins does Ross have?

**18.**

Josh has 88 fewer coins than May.

May has 95 coins.

How many coins does Josh have?

◯ Draw ▭. Then fill in the numbers you know.
◯ Add or subtract to find the number you do not know.
◯ Write the answer in a word sentence.

**19.**

Mindy drew 5 stars. Jon drew 4 more stars than Mindy. How many stars did Jon draw?

**20.**

Alex ran around the track 10 times. Billy ran around the track 6 fewer times than Alex. How many times did Billy run around the track?

**21.**

Mona peeled 24 oranges. Eddy peeled 5 more oranges than Mona. How many oranges did Eddy peel?

**22.**

Ted made 38 snowballs. Cathy made 12 fewer snowballs than Ted. How many snowballs did Cathy make?

**23.**

Sandy wrote 55 text messages. Bob wrote 10 more messages than Sandy. How many messages did Bob write?

☐ Draw a line to show where the first step ends.

☐ Use the comparing picture to find the smaller number.

☐ Use the part-whole picture to find the total.

**1.** Zara counted 5 trucks. Greg counted 3 fewer trucks than Zara. | How many trucks did they count altogether?

Zara _____

Greg _____

$5 + 2 = 7$

**2.** Jack counted 8 turtles. Anna counted 6 fewer turtles than Jack. How many turtles did they count altogether?

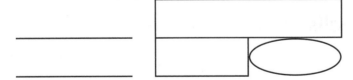

☐ Draw a line to show where the first step ends.

☐ Use the comparing picture to find the smaller number.

☐ Use the part-whole picture to find the total.

**3.**

Cameron sang 10 songs. Grace sang 3 fewer songs than Cameron. How many songs did they sing altogether?

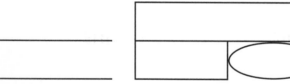

_____

**4.**

Josh wrote 9 stories. Fred wrote 8 fewer stories than Josh. How many stories did they write altogether?

**5. BONUS**

Beth read 61 comics. Carlos read 35 fewer comics than Beth. How many comics did they read altogether?

# OA2-61 More Word Problems

○ Find the answer.

1.
> Carl picked 49 fewer blueberries than Rani. Rani picked 54 blueberries. How many blueberries did Carl pick?

2.
> There were 18 seals on the shore. 8 swam away. Then 11 came back to the shore. How many seals are there now?

3.
> There were 15 used bicycles and some new bicycles in the store. There were 53 bicycles altogether. Then the store got 2 more new bicycles. How many new bicycles are there now?

4.
> Mary walked up 58 steps. Then she walked up 27 steps. How many steps did she walk up altogether?

5.
> Jill ate 23 more raisins than Amit. Amit ate 67 raisins. How many raisins did Jill eat?

6.
> A small bird hopped 20 meters. A large bird hopped 10 fewer meters. How many meters did the large bird hop?

7.
> Dan picked 38 carrots. Then he picked some more carrots. Now there are 67 carrots. How many more carrots did Dan pick?

# NBT2-41 Counting by Tens and Hundreds (I)

◯ What number do the tens blocks show?

**1.**
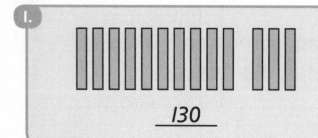

_130_

**2.**
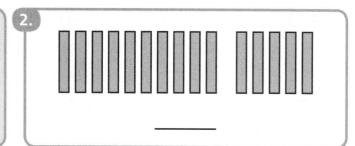

_____

**3.**
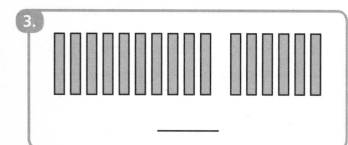

_____

**4.**
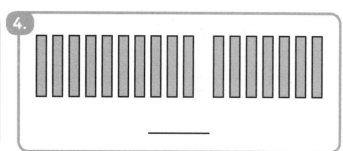

_____

◯ Draw tens to show the number.

**5.**
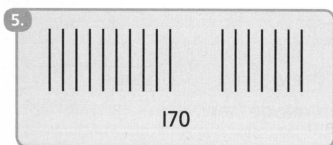

170

**6.**

120

◯ Add.

**7.**
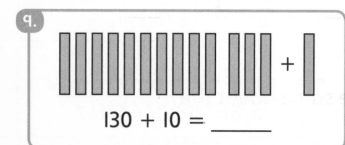

120 + 10 = _130_

**8.**
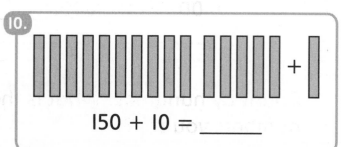

140 + 10 = _____

**9.**

130 + 10 = _____

**10.**
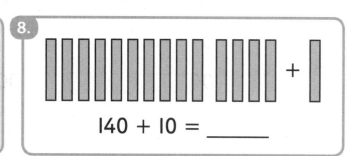

150 + 10 = _____

◯ Add.

**11.** $120 + 10 = \underline{130}$

**12.** $150 + 10 = \underline{\phantom{xxx}}$

**13.** $180 + 10 = \underline{\phantom{xxx}}$

**14.** $270 + 10 = \underline{\phantom{xxx}}$

**15.** $460 + 10 = \underline{\phantom{xxx}}$

**16.** $690 + 10 = \underline{\phantom{xxx}}$

◯ Count on by tens.

**17.** 10, 20, 30, $\underline{40}$ , $\underline{50}$

**18.** 70, 80, 90, $\underline{\phantom{xxx}}$ , $\underline{\phantom{xxx}}$

**19.** 200, 210, 220, $\underline{\phantom{xxx}}$ , $\underline{\phantom{xxx}}$

**20.** 440, 450, 460, $\underline{\phantom{xxx}}$ , $\underline{\phantom{xxx}}$

**21.** 750, 760, 770, $\underline{\phantom{xxx}}$ , $\underline{\phantom{xxx}}$

**22.** 960, 970, 980, $\underline{\phantom{xxx}}$ , $\underline{\phantom{xxx}}$

◯ Count on by hundreds to find the missing numbers.

**23.** 100, 200, 300, $\underline{400}$ , 500, 600, $\underline{700}$ , 800, 900

**24.** 200, $\underline{\phantom{xxx}}$ , $\underline{\phantom{xxx}}$ , 500, 600, 700, $\underline{\phantom{xxx}}$ , 900

**25. BONUS**

3400, 3500, 3600, $\underline{\phantom{xxxx}}$ , $\underline{\phantom{xxxx}}$ , $\underline{\phantom{xxxx}}$

**26. BONUS**

Count by hundreds. What is the same about the numbers you say?

# NBT2-42 Counting by Tens and Hundreds (2)

⬜ Write the number the blocks show.

**1.**
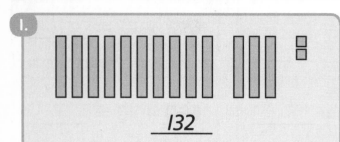

_132_

**2.**
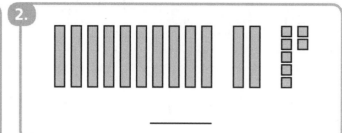

_____

⬜ Draw tens to show the number.

**3.**

174

**4.**
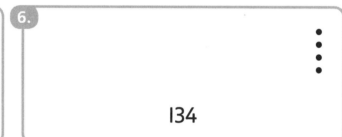

114

**5.**

144

**6.**
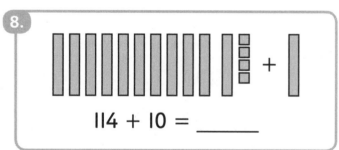

134

⬜ Add.

**7.**
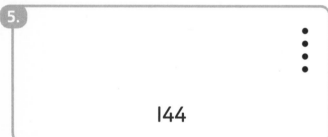

124 + 10 = _134_

**8.**

114 + 10 = _____

**9.**
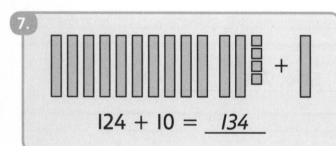

143 + 10 = _____

**10.**
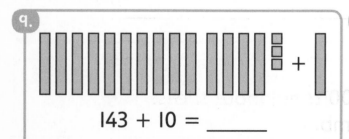

132 + 10 = _____

☐ Add.

**11.** 126 + 10 = __136__

**12.** 185 + 10 = _____

**13.** 698 + 10 = _____

**14.** 491 + 10 = _____

**15.** 343 + 10 = _____

**16.** 502 + 10 = _____

☐ Add 10 each time.

**17.** 15, 25, 35, __45__, __55__

**18.** 73, 83, 93, _____, _____

**19.** 108, 118, 128, _____, _____

**20.** 651, 661, 671, _____, _____

☐ Add 100 each time.

**21.** 110, 210, 310, __410__, __510__, __610__

**22.** 406, 506, 606, _____, _____, _____

**23.** 352, 452, 552, _____, _____, _____

**24. BONUS** 2045, 2145, 2245, _____, _____, _____

**25. BONUS**

Start at 27 and add 100. Add 100 three more times.
What is the same in all the numbers?

# NBT2-43 Counting by Fives

◯ Start at 5. Add fives and shade the answer each time.

◯ Fill in the blanks.

**1.**

| 1 | 2 | 3 | 4 | 5 | 6 | 7 | 8 | 9 | 10 |
|---|---|---|---|---|---|---|---|---|----|
| 11 | 12 | 13 | 14 | 15 | 16 | 17 | 18 | 19 | 20 |
| 21 | 22 | 23 | 24 | 25 | 26 | 27 | 28 | 29 | 30 |
| 31 | 32 | 33 | 34 | 35 | 36 | 37 | 38 | 39 | 40 |

The ones digit of the shaded numbers is _____ or _____ .

◯ Count on by fives.

**2.**

5, 10, 15, __20__ , __25__ , __30__

**3.**

60, 65, 70, _____, _____, _____

**4.**

105, 110, 115, _____, _____

**5.**

340, 345, 350, _____, _____

◯ Circle the numbers you say if you count on by fives starting at 5.

**6.**

1        53              37              55

            5

            66              75

70                    61

46                  95              16

        82                  20

◯ **BONUS:** How did you know which numbers to circle?

**7.**

_____

○ Start at 3. Add fives and shade the answer each time.
○ Fill in the blanks.

**8.**

| 1 | 2 | 3 | 4 | 5 | 6 | 7 | 8 | 9 | 10 |
|---|---|---|---|---|---|---|---|---|---|
| 11 | 12 | 13 | 14 | 15 | 16 | 17 | 18 | 19 | 20 |
| 21 | 22 | 23 | 24 | 25 | 26 | 27 | 28 | 29 | 30 |
| 31 | 32 | 33 | 34 | 35 | 36 | 37 | 38 | 39 | 40 |

The ones digit of the shaded numbers is _____ or _____.

○ Add 5 each time.

**9.**

2, 7, 12, __17__, __22__, __27__

**10.**

21, 26, 31, _____, _____, _____

**11.**

38, 43, 48, _____, _____, _____

**12.**

54, 59, 64, _____, _____, _____

○ Circle the numbers you say if you add fives starting at 1.

**13.**

56      37

1      5      21

61

70      75

82      16

66

45      95      20

**14. BONUS**

Lily starts at 4 and adds fives. What are the ones digits of all the numbers?

# NBT2-44 Fluency with 10 and 100

☐ Add.

**1.**
$120 + 100 =$ _____

$120 + 10 =$ _____

$120 + 1 =$ _____

**2.**
$340 + 100 =$ _____

$340 + 10 =$ _____

$340 + 1 =$ _____

**3.**
$653 + 100 =$ _____

$653 + 10 =$ _____

$653 + 1 =$ _____

**4.**
$470 + 100 =$ _____

**5.**
$280 + 10 =$ _____

**6.**
$650 + 1 =$ _____

**7.**
$853 + 100 =$ _____

**8.**
$742 + 10 =$ _____

**9.**
$478 + 1 =$ _____

**10.**
$555 + 10 =$ _____

**11.**
$555 + 1 =$ _____

**12.**
$555 + 100 =$ _____

**13.**
$777 + 1 =$ _____

**14.**
$777 + 100 =$ _____

**15.**
$777 + 10 =$ _____

☐ Write the missing number.

**16.**
$220 = 200 +$ _____

**17.**
$202 = 200 +$ _____

**18.**
$404 = 400 +$ _____

**19.**
$305 = 300 +$ _____

**20.**
$260 =$ _____ $+ 60$

**21.**
$620 = 600 +$ _____

**22. BONUS**

Jim says $100 + 1 = 110$. What mistake did he make?

_____

☐ Subtract.

**23.** $170 - 10 = \underline{\textit{160}}$

**24.** $140 - 10 = \underline{\hspace{1cm}}$

**25.** $295 - 10 = \underline{\hspace{1cm}}$

**26.** $614 - 10 = \underline{\hspace{1cm}}$

**27.** $458 - 10 = \underline{\hspace{1cm}}$

**28.** $806 - 10 = \underline{\hspace{1cm}}$

☐ Subtract 10 each time.

**29.** 90, 80, 70, $\underline{\textit{60}}$, $\underline{\textit{50}}$

**30.** 60, 50, 40, $\underline{\hspace{1cm}}$, $\underline{\hspace{1cm}}$

**31.** 110, 100, 90, $\underline{\hspace{1cm}}$, $\underline{\hspace{1cm}}$

**32.** 257, 247, 237, $\underline{\hspace{1cm}}$, $\underline{\hspace{1cm}}$

☐ Subtract.

**33.** $200 - 100 = \underline{\textit{100}}$

**34.** $300 - 100 = \underline{\hspace{1cm}}$

**35.** $540 - 100 = \underline{\hspace{1cm}}$

**36.** $719 - 100 = \underline{\hspace{1cm}}$

**37.** $956 - 100 = \underline{\hspace{1cm}}$

**38.** $802 - 100 = \underline{\hspace{1cm}}$

☐ Subtract 100 each time.

**39.** 900, 800, 700, $\underline{\textit{600}}$, $\underline{\textit{500}}$

**40.** 600, 500, 400, $\underline{\hspace{1cm}}$, $\underline{\hspace{1cm}}$

**41.** 850, 750, 650, $\underline{\hspace{1cm}}$, $\underline{\hspace{1cm}}$

**42.** 534, 434, 334, $\underline{\hspace{1cm}}$, $\underline{\hspace{1cm}}$

The boxes have the same number.

⬜ Write the numbers to make the total.

**1.**

```
     ┌───┐
     │ 7 │
     ├───┤
  +  │ 7 │
     └───┘
  ─────────
     14
```

**2.**

```
     ┌───┐
     │   │
     ├───┤
  +  │   │
     └───┘
  ─────────
      6
```

**3.**

```
     ┌───┐
     │   │
     ├───┤
  +  │   │
     └───┘
  ─────────
     16
```

**4.**

```
    1  ┌───┐
       │   │
       ├───┤
  + 2  │   │
       └───┘
  ─────────
    3    4
```

**5.**

```
    1  ┌───┐
       │   │
       ├───┤
  + 2  │   │
       └───┘
  ─────────
    4    4
```

**6.**

```
    3  ┌───┐
       │   │
       ├───┤
  + 5  │   │
       └───┘
  ─────────
    8    2
```

**7.**

```
   1  3  ┌───┐
         │   │
      ┌──┼───┤
  + 2 │  │   │  5
      └──┴───┘
  ──────────────
   3  5  7
```

**8.**

```
   2  ┌───┐  1
      │   │
      ├───┤
  + 3 │   │  7
      └───┘
  ──────────────
   5  2  8
```

**9.**

```
   1  ┌───┐  5
      │   │
      ├───┤
  + 6 │   │  8
      └───┘
  ──────────────
   7  3  3
```

**10.**

In the 2014 Winter Olympics, the United States won 9 gold, 7 silver, and 12 bronze medals. Canada won 10 gold, 10 silver, and 5 bronze medals.

How many more gold medals did Canada win? _____

How many medals did the United States win? _____

How many medals did Canada win? _____

How many more medals did the United States win altogether?

○ Who am I?

**11.**

You say me when you count by fives. I am greater than 15 but less than 22.

5    10    15    (20)    25

**12.**

I am a two-digit number. You say me when you count by fives. My tens digit is 5. I am less than 53.

**13.**

I am a three-digit number. You say me when you start at 123 and add hundreds. My digits add to 9.

**14. BONUS**

I am a two-digit number. You say me when you start at 43 and add fives. My ones digit is 3 more than my tens digit.

**15. BONUS**

I am a three-digit number. My hundreds digit is greater than my tens digit. My tens digit is greater than my ones digit. My digits add to 6.

# MD2-I7 Measuring in Inches

We measure length, width, or height of small objects in **inches**.

You start at zero and count jumps to measure in inches.

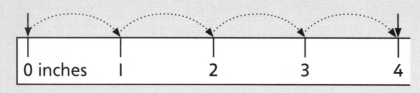

4 jumps = 4 inches

◯ Measure the distance between the arrows by counting jumps.

1.

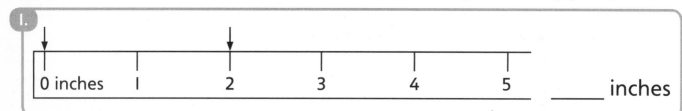

_____ inches

2.

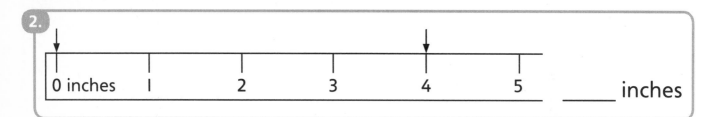

_____ inches

3.

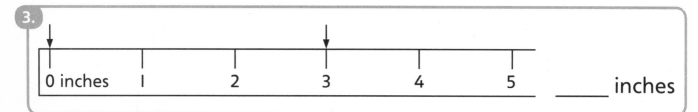

_____ inches

4.

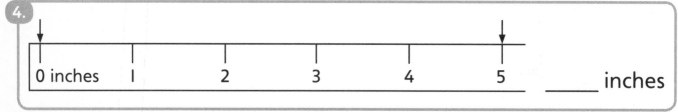

_____ inches

5. BONUS

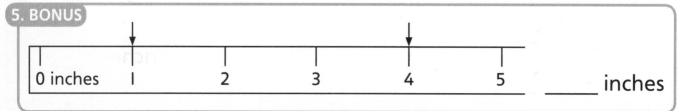

_____ inches

☐ How long is the line or object?

**6.**
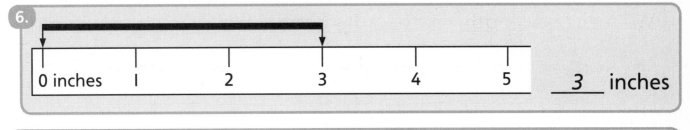
_3_ inches

**7.**
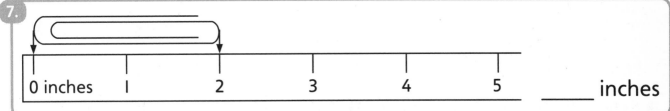
_____ inches

**8.**
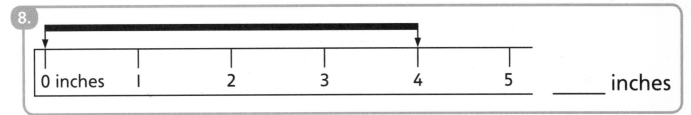
_____ inches

**9.**
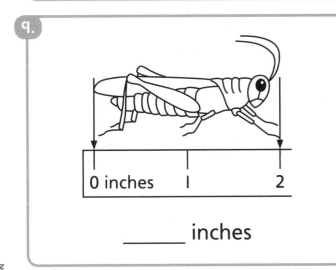
_____ inches

**10.**
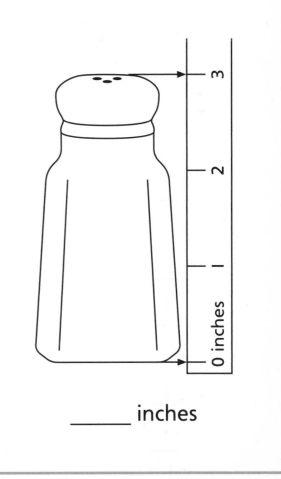
_____ inches

**11.**
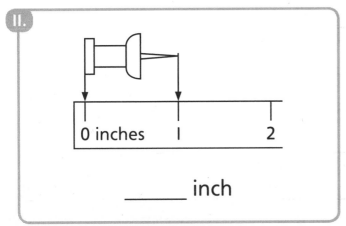
_____ inch

○ About how long is the object?

12.
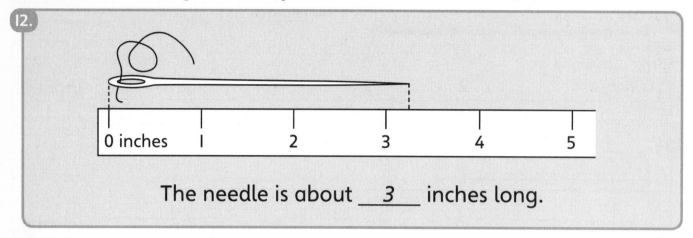

The needle is about ___3___ inches long.

13.
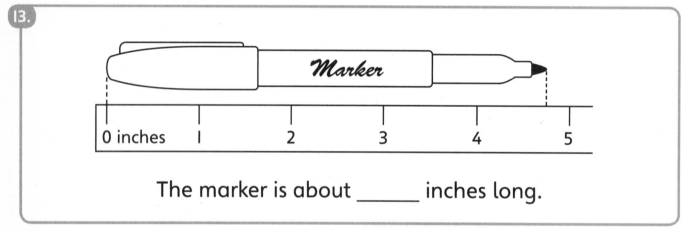

The marker is about _____ inches long.

14.
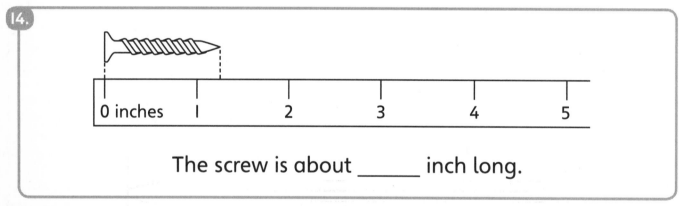

The screw is about _____ inch long.

15.
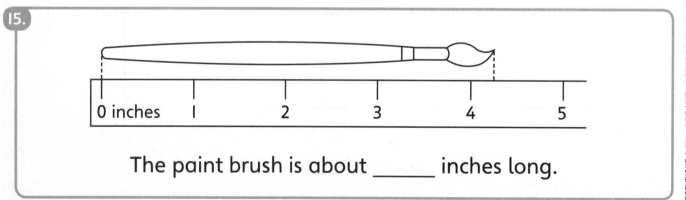

The paint brush is about _____ inches long.

Measurement and Data 2-17

A pattern block square is 1 inch long.
☐ About how many inches long is the object?

**16.**

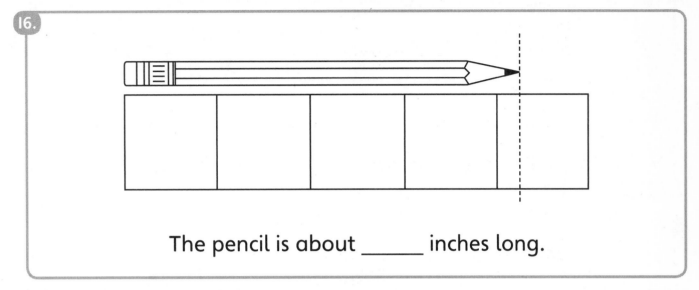

The pencil is about _____ inches long.

**17.**

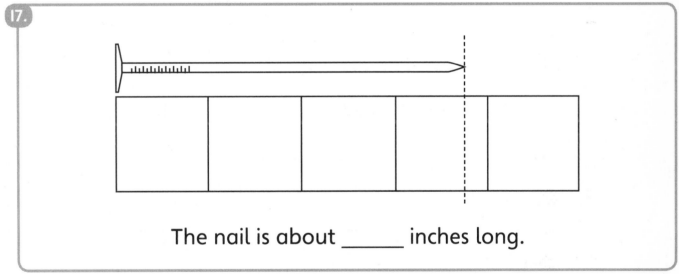

The nail is about _____ inches long.

**18.**

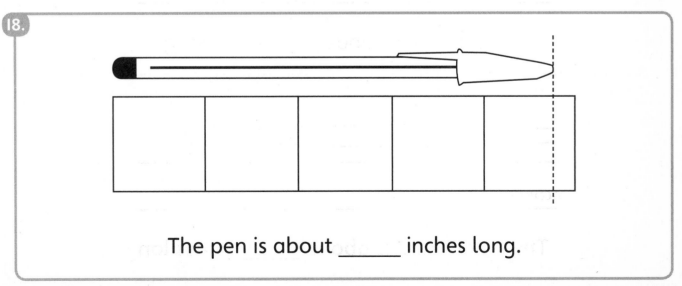

The pen is about _____ inches long.

# MD2-18 Estimating and Measuring in Inches

Two fingers are about 1 inch wide.

about 1 inch

0 inches    1        2

☐ Use your fingers to estimate the length to the nearest inch.

☐ Measure the length in inches.

**1.**

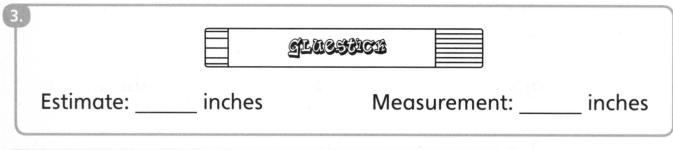

Estimate: _____ inches      Measurement: _____ inches

**2.**

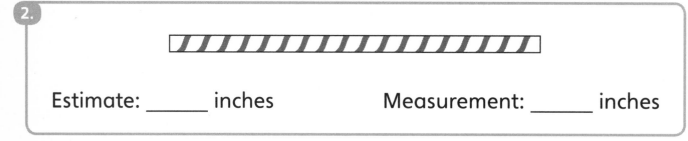

Estimate: _____ inches      Measurement: _____ inches

**3.**

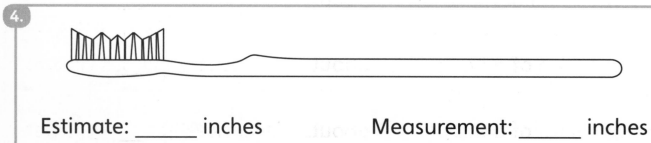

Estimate: _____ inches      Measurement: _____ inches

**4.**

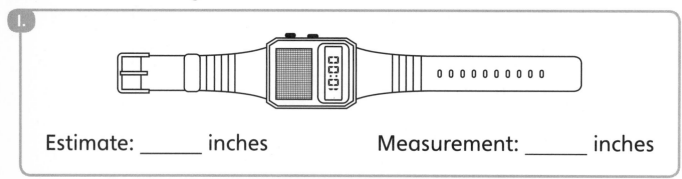

Estimate: _____ inches      Measurement: _____ inches

## What is the best estimate for the bottom object?
☐ Write a ✓.

**5.**

6 inches

☐ 1 inch
☐ 3 inches
☐ 7 inches

**6.**

9 inches

☐ 3 inches
☐ 10 inches
☐ 14 inches

**7.**

13 inches

☐ 10 inches
☐ 14 inches
☐ 18 inches

☐ Estimate to the nearest inch.
☐ Measure to the nearest inch.

**8.**

| Object | Estimate | Measurement |
|---|---|---|
| The length of a pen | about _____ inches | _____ inches |
| The width of a book | about _____ inches | _____ inches |
| The width of a desk | about _____ inches | _____ inches |

# MD2-19 Measuring in Different Units

How long is the line?

**1.**

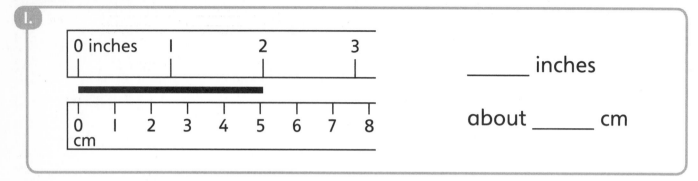

_____ inches

about _____ cm

**2.**

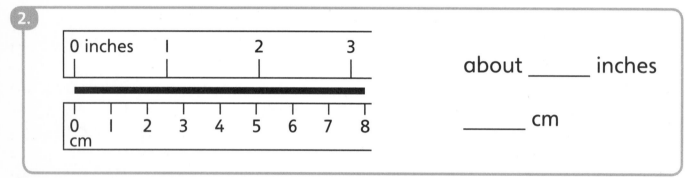

about _____ inches

_____ cm

**3.**

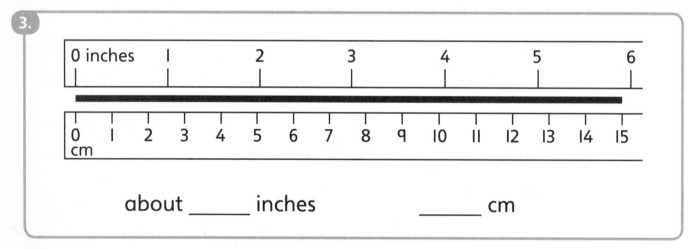

about _____ inches          _____ cm

**4. BONUS**

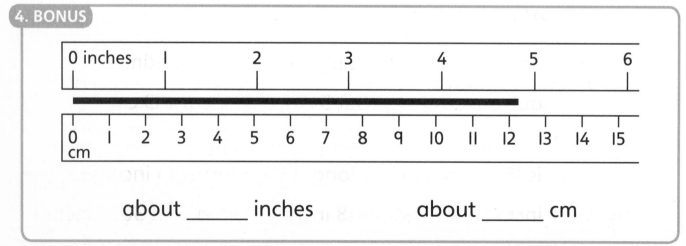

about _____ inches          about _____ cm

◯ Measure the object in centimeters (cm) and in inches.

**5.**

_____ cm

about _____ inches

**6.**

_____ cm

about _____ inches

**7.**

about _____ inches      about _____ cm

**8.**

about _____ inches      about _____ cm

◯ Write **longer** or **shorter**.

**9.**

I inch is _____ than I cm.

I cm is _____ than I inch.

◯ Circle the correct length.

**10.**

A pen is 5 inches long. How long is the pen in centimeters?

about 2 cm       about 5 cm       about 13 cm

**11.**

A marker is 18 cm long. How long is the marker in inches?

about 7 inches       about 18 inches       about 30 inches

# MD2-20 Estimating and Measuring in Feet

From your fingertips to your elbow is about I **foot** long.

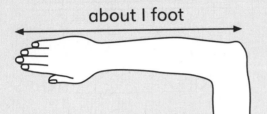

about I foot

You can write **ft** for foot.

A foot is 12 inches. You can measure feet using a 12-inch ruler.

☐ Estimate the **width** to the nearest foot.

☐ Measure to the nearest foot.

| Object | Estimate | Measurement |
|---|---|---|
| A table | about _____ ft | _____ ft |
| A window | about _____ ft | _____ ft |
| A door | about _____ ft | _____ ft |
| A blackboard | about _____ ft | _____ ft |
| A bookshelf | about _____ ft | _____ ft |
| A desk | about _____ ft | _____ ft |
| A poster | about _____ ft | _____ ft |
| The seat of a chair | about _____ ft | _____ ft |

**Measurement and Data 2-20**

# MD2-21 Estimating and Measuring in Feet and Yards

☐ Circle the things that are **longer** than 1 foot.

☐ Circle the things that are **shorter** than 1 foot.

☐ Circle the things that are about 1 foot long.

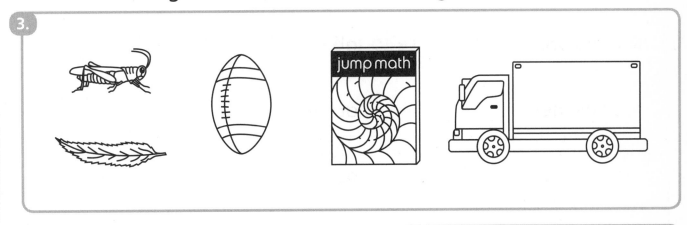

**4.** Name 3 objects in your classroom that are about 1 foot long.

_____, _____, _____

A **yard** is 3 feet long. You can write **yd** for yard.

A **yardstick** is 3 feet long.    A big step is about I yard long.

I yard

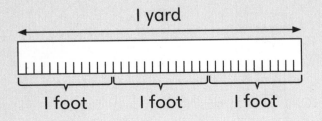

| I foot | I foot | I foot |

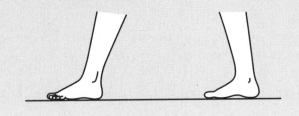

☐ Complete the table.

**5.**

| Yards | Feet |
|---|---|
| I yard | *3 feet* |
| 2 yards | *6 feet* |
| 3 yards | |
| 4 yards | |

☐ Use big steps to estimate to the nearest yard.
☐ Use a yardstick to measure to the nearest yard.

**6.**

| Object | Estimate | Measurement |
|---|---|---|
| The width of your classroom | about _____ yd | _____ yd |
| The length of your classroom | about _____ yd | _____ yd |
| The length of a hallway | about _____ yd | _____ yd |

# MD2-22 Choosing Tools and Units

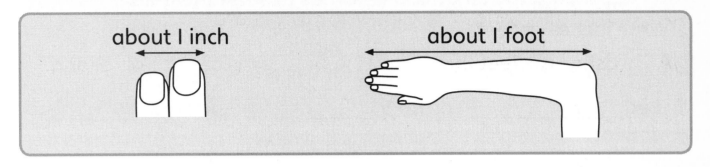

about I inch

about I foot

☐ Circle the objects that are more than I foot long.

☐ Cross out the objects that are less than I inch long.

**1.**

Which unit would you use to measure length?

☐ Circle **inches** or **feet**.

**2.**

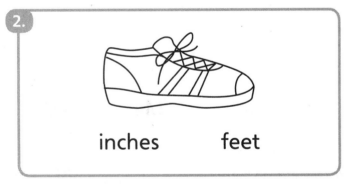

inches    feet

**3.**

inches    feet

**4.**

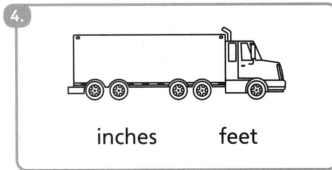

inches    feet

**5.**

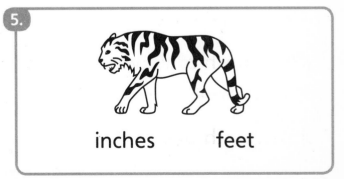

inches    feet

Kim measured some lengths. She forgot to write the units.
◯ Write **inches** or **feet**.

**6.**

bed 7 _____    brush 5 _____    piano 5 _____

You can write **in** for inch.
◯ Fill in the blank with **in** or **ft**.

**7.**

A bike is 5 _____ long

**8.**

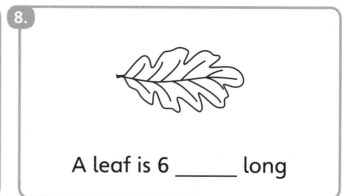

A leaf is 6 _____ long

**9. BONUS**

Niagara Falls is

160 _____ tall.

**10. BONUS**

A hamster is

4 _____ long.

You can use a 12-inch ruler to measure lengths less than 1 foot.
You can use a yardstick for lengths less than 3 feet.
You can use a measuring tape for lengths greater than 3 feet.

◯ What tool would you use to measure the length?
◯ Explain why.

**11.**

sunglasses

**12.**

swimming pool

**13.**

desk

# MD2-23 Word Problems

◯ Draw a line to show the length of the hamster.

◯ Make the line longer to show how much the hamster grows.

**1.**

A hamster is 3 inches long. It grows 2 inches.

How long is the hamster now?

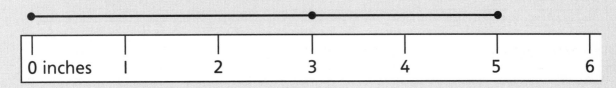

The hamster is ___5___ inches long.

**2.**

A hamster is 4 inches long. It grows 2 inches.

How long is the hamster now?

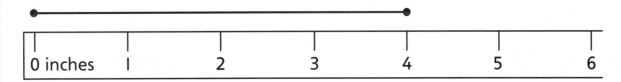

The hamster is _____ inches long.

**3.**

A hamster is 3 inches long. It grows 1 inch.

How long is the hamster now?

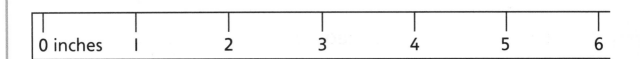

The hamster is _____ inches long.

☐ Draw a line to show the length.
☐ Draw an arrow to show how much to take away.

**4.**

Anna draws a line 6 inches long. She erases 2 inches.
How long is the line now?

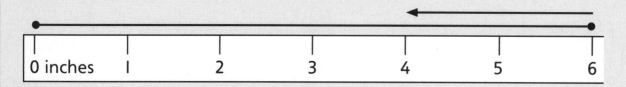

The line is __4__ inches long.

**5.**

Rick draws a line 5 inches long. He erases 3 inches.
How long is the line now?

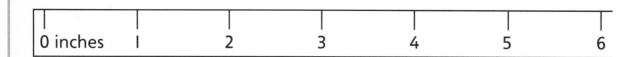

The line is _____ inches long.

**6.**

Mona's string is 6 inches long. She cuts off 4 inches.
How long is the string now?

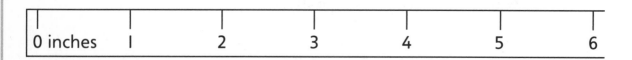

The string is _____ inches long.

⬭ Match the problem with one of the pictures.

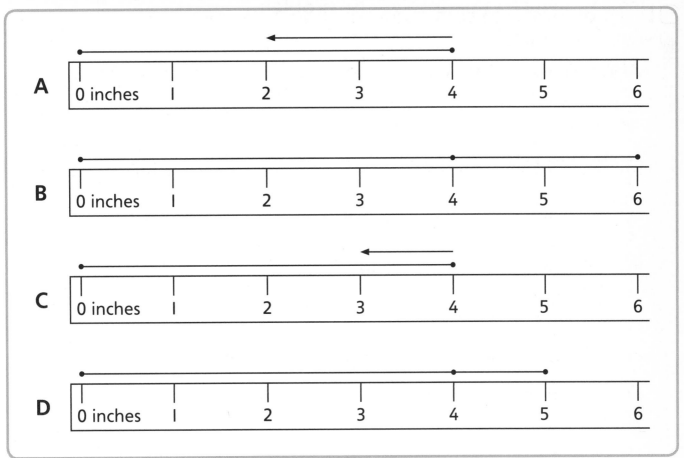

7.
A straw is 4 inches long. Jon cuts off 2 inches. __A__

8.
A line is 4 inches long. Cam adds 1 inch. _____

9.
A carrot is 4 inches long. Zara eats 1 inch. _____

10.
A worm is 4 inches long. It grows 2 inches. _____

# MD2-24 Word Problems and Equations (I)

☐ Draw a picture to help solve the problem.
☐ Write the equation.
☐ Write the answer.

**1.**

Mary draws a line 6 inches long. She erases 3 inches.
How long is the line now?

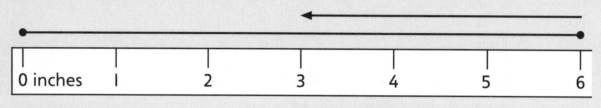

$\underline{\quad 6 - 3 = 3 \quad}$          The line is __3__ inches long.

**2.**

A crayon is 4 inches long. Ross breaks off 2 inches.
How long is the crayon now?

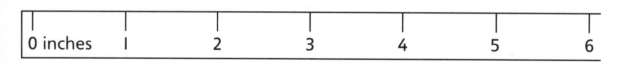

$\underline{\hspace{4cm}}$          The crayon is _____ inches long.

**3.**

A banana is 6 inches long. A monkey eats 5 inches.
How long is the banana now?

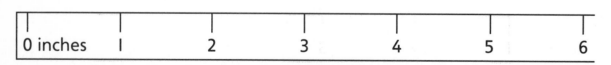

$\underline{\hspace{4cm}}$          The banana is _____ inch long.

☐ Draw a picture.
☐ Write the equation.
☐ Write the answer.

**4.**

A lizard is 2 inches long. It grows 3 inches.
How long is it now?

| 0 inches | 1 | 2 | 3 | 4 | 5 | 6 |

_____          The lizard is _____ inches long.

**5.**

A tomato plant is 5 inches tall. It grows 1 inch.
How tall is it now?

| 0 inches | 1 | 2 | 3 | 4 | 5 | 6 |

_____          _____.

**6.**

A turtle is 4 inches long. It grows 2 inches.
How long is it now?

| 0 inches | 1 | 2 | 3 | 4 | 5 | 6 |

_____          _____.

# MD2-25 Word Problems and Equations (2)

○ Write the numbers you know in the blanks.
○ Draw ☐ for the number you do not know.
○ Use the part-whole picture to find the number you do not know.

**1.**

Carl walked 50 yards to the store.
Then he walked to to the library.
Altogether he walked 90 yards.
How far did he walk from the store to the library?

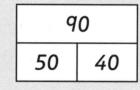

<u>  50  </u>    +    <u>|40|</u>    =    <u>  90  </u>

yards to store     yards to library     total yards

**2.**

Kathy walked 60 yards in the morning.
She walked 70 yards in the afternoon.
How far did she walk altogether?

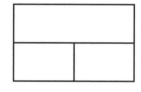

<u>      </u>    +    <u>      </u>    =    <u>      </u>

yards in morning     yards in afternoon     total yards

**3.**

Ravi walked 60 feet to his classroom.
Then he walked to the gym.
Altogether he walked 80 feet.
How far did he walk from his classroom to the gym?

<u>      </u>    +    <u>      </u>    =    <u>      </u>

feet to classroom     feet to gym     total feet

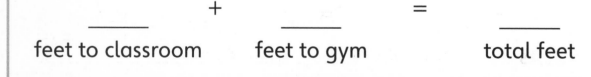

○ Write the numbers you know in the blanks.
○ Draw ☐ for the number you do not know.
○ Use the part-whole picture to find the number you do not know.

**4.**

Mindy drew a line 20 inches long. She erased part of the line. Then her line was 5 inches long. How much did she erase?

| 20 | |
|---|---|
| 5 | |

_20_ — _____ = _5_
total inches    inches erased    inches left

**5.**

A chain of paper clips was 18 inches long. Ted took off some paper clips. Then the chain was 12 inches long. How much did he take away?

| | |
|---|---|
| | |

_____ — _____ = _____
total inches    inches taken away    inches left

**6.**

A string was 47 inches long. Tina cut off 13 inches. How many inches are left?

| | |
|---|---|
| | |

_____ — _____ = _____
total inches    inches cut off    inches left

# MD2-26 Two-Step Word Problems with Length

☐ Write the numbers in the blanks.

☐ Write + or − in the circles.

☐ Find the answer.

**1.**

In Alaska it snowed 3 feet in December, 7 feet in January, and 9 feet in February. How many feet did it snow altogether?

_____ ◯ _____ ◯ _____ = _____

**2.**

Jane's plant is 37 inches tall. She cuts off 7 inches. Then it grows 9 more inches. How many inches tall is it now?

_____ ◯ _____ ◯ _____ = _____

☐ Subtract using the numbers you know.

☐ Use the answer for the subtraction to complete the question.

**3.**

Rachel had 4 feet of red ribbon and some green ribbon. She had 14 feet of ribbon altogether. Then she bought 7 more feet of green ribbon. How many feet of green ribbon does she have now?

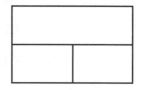

_____ − _____ = _____          _____ + _____ = _____

Rachel has _____ feet of green ribbon.

☐ Draw a line to show where the first step ends.
☐ Use the comparing picture to find the smaller number.
☐ Use the part-whole picture to find the total.

**4.**

Lily dug a hole that is 23 inches. Jason dug a hole that is 3 inches shorter than Lily's. How many inches did they dig altogether?

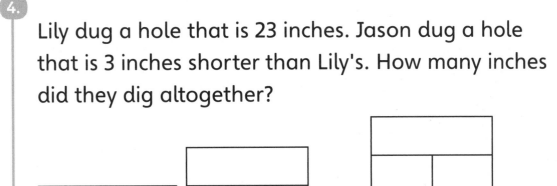

_____ + _____ = _____

☐ Find the answer.

**5.**

Lee had 6 feet of blue ribbon and some red ribbon. She had 9 feet of ribbon altogether. Then she bought 2 more feet of red ribbon. How many feet of red ribbon does she have now?

**6.**

Amit measured a room that is 35 meters long.
Grace measured a room that is 4 fewer meters long.
How many meters did they measure altogether?

**7.**

Fred had 13 inches of red string and some blue string.
He had 18 inches of string altogether. Then his teacher gave him 10 more inches of blue string. How many inches of blue string does he have now?

# MD2-27 Line Plots

Jin measured some fish and started a line plot. For each fish, he drew an X on the line plot.

☐ Finish the line plot.

**1.**

| Fish | Length |
|------|--------|
| A | ~~6 inches~~ |
| B | ~~4 inches~~ |
| C | ~~4 inches~~ |
| D | 5 inches |
| E | 6 inches |

**Lengths of Fish**

C ⟶ X
B ⟶ X        X ⟵ A

4    5    6

Length (inches)

☐ Make a line plot. Cross out each length as you go.

**2.**

| Dog | Height |
|------|--------|
| Collie | 2 feet |
| Great Dane | 4 feet |
| Husky | 3 feet |
| Poodle | 2 feet |
| Pug | ~~1 foot~~ |
| Sheepdog | 2 feet |
| Dalmation | 3 feet |
| Beagle | 1 foot |

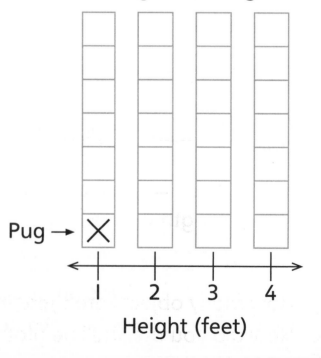

**Heights of Dogs**

Pug ⟶ X

1    2    3    4

Height (feet)

☐ Measure the objects.

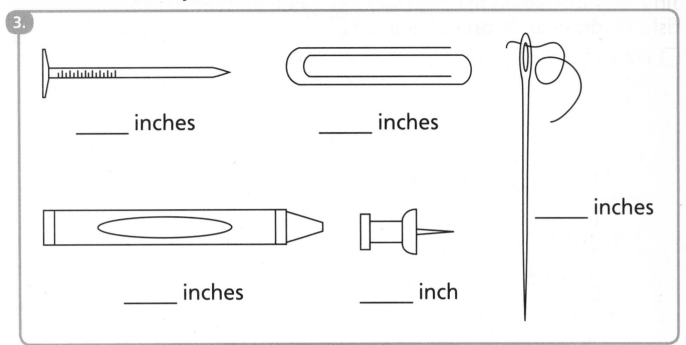

**3.**

_____ inches

_____ inches

_____ inches

_____ inches

_____ inch

_____ inches

☐ Use the lengths of the objects above to make a line plot.
☐ Answer the questions.

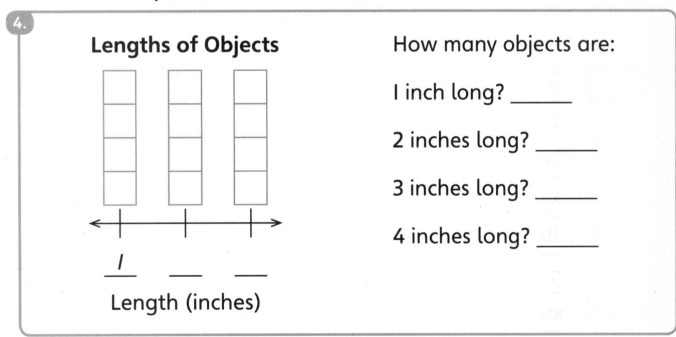

**4.**

**Lengths of Objects**

Length (inches)

How many objects are:

1 inch long? _____

2 inches long? _____

3 inches long? _____

4 inches long? _____

**5.**

How many objects are there in total?
How did you use the line plot to find the total?

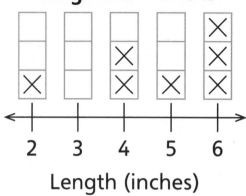

**Lengths of Pencils**

Length (inches)

☐ Answer the questions about the line plot.

**6.**

How many pencils are 2 inches long? _____

**7.**

How many pencils are 3 inches long? _____

**8.**

How many pencils are 4 inches long? _____

**9.**

What is the most common length of pencil? _____

**10.**

How many pencils are there altogether? _____

**11.**

How many more pencils are 6 inches than 4 inches? _____

# MD2-28 Problems and Puzzles

Choose the best estimate for the bottom object.
☐ Write a ✓.

**1.**

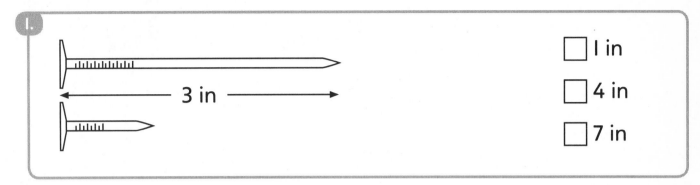

3 in

☐ 1 in
☐ 4 in
☐ 7 in

**2.**

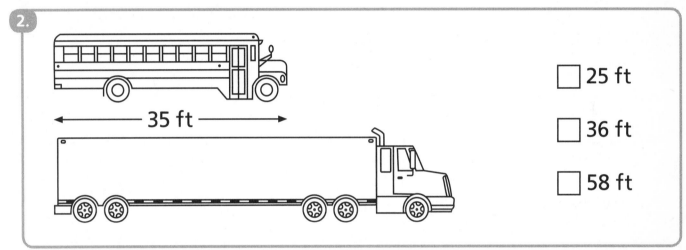

35 ft

☐ 25 ft
☐ 36 ft
☐ 58 ft

**3.** A playground is 98 yards long. A school is 80 yards long. How much longer is the playground?

**4.** Zack has a scooter that is 30 inches long. Sun has a skateboard that is 3 inches longer. How long is the skateboard?

**5.** One truck is 32 feet long. How long are two trucks?

**6.** One ladder is 14 feet long. How long are three ladders?

| Animal | Height |
| --- | --- |
| Kangaroo | 5 feet |
| Penguin | 3 feet |
| Giraffe | 19 feet |
| Moose | 7 feet |

◻ Use the table to order the animals from shortest to tallest.

**7.**

1. _____    2. _____

3. _____    4. _____

◻ Use the table to answer the questions.

**8.**

How much taller than a penguin is a giraffe? _____

**9.**

How much taller than a kangaroo is a moose? _____

**10.**

How much shorter than a kangaroo is a penguin? _____

**11.**

How much shorter than a giraffe is a kangaroo? _____

This is a clock face.

The numbers start at 1 and end at 12.

⬜ Fill in the missing 3, 6, 9, or 12.

| 1. | 2. | 3. | 4. |
|---|---|---|---|

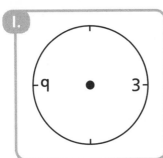 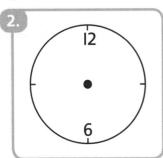 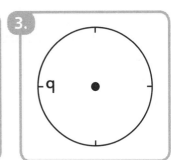 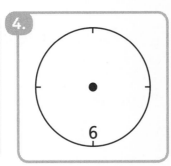

⬜ Fill in the missing numbers. Start with 3, 6, 9, and 12.

**5.**

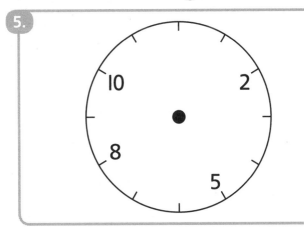

**6.**

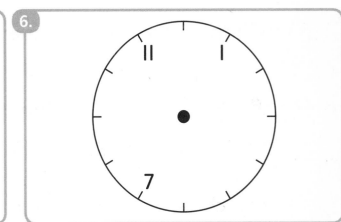

**7.**

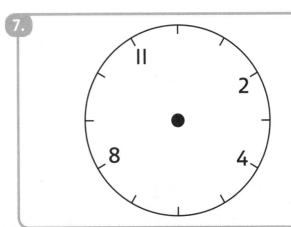

**8.**

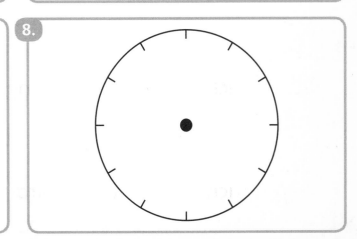

The **hour hand** is shorter.

hour hand

The hour hand is pointing **at the 3**.

☐ Where is the hour hand pointing?

**9.**

at the _____

**10.**

at the _____

**11.**

at the _____

**12.**

at the _____

**13.**

at the _____

**14.**

at the _____

## Analog Clock

## Digital Clock

It is **9 o'clock** or **9:00**.

⬜ Write the time two ways.

**15.**

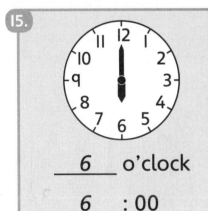

<u>  6  </u> o'clock

<u>  6  </u> : 00

**16.**

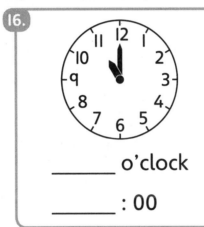

_____ o'clock

_____ : 00

**17.**

_____ o'clock

_____ : 00

**18.**

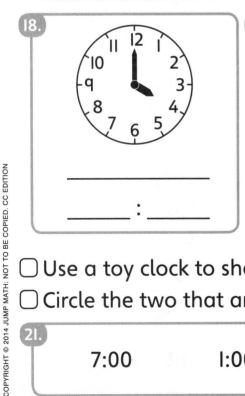

_____

_____ : _____

**19.**

_____

_____ : _____

**20.**

_____

_____ : _____

⬜ Use a toy clock to show these times.
⬜ Circle the two that are the same.

**21.**

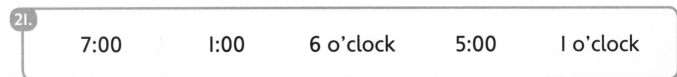

7:00          1:00          6 o'clock          5:00          1 o'clock

☐ Draw the hour hand.

**22.**

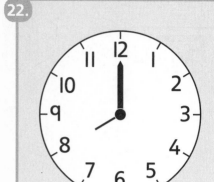

8 o'clock

**23.**

7 o'clock

**24.**

3 o'clock

**25.**

6:00

**26.**

4:00

**27.**

1:00

**28.**

Kim starts playing soccer at the time on the clock. She plays for I hour. When does she stop playing?

_____

_____

_____

# MD2-30 The Minute Hand

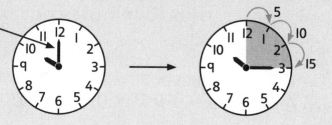

The **minute hand** is longer.

minute hand

Count by 5s.

It is **15** minutes after 10:00.

☐ How many minutes is it after 10:00?

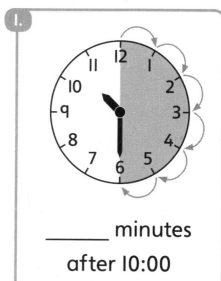

**1.**

_____ minutes

after 10:00

**2.**

_____ minutes

after 10:00

**3.**

_____ minutes

after 10:00

☐ Write how many minutes after the hour.

**4.**

__15__ minutes

after 7:00

**5.**

_____ minutes

after 8:00

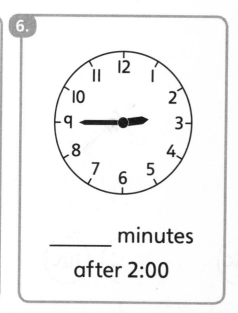

**6.**

_____ minutes

after 2:00

It is 10 minutes past 4:00.

☐ : 10 ← Write the minutes past here.

☐ Write the number of minutes past the hour.

**7.**

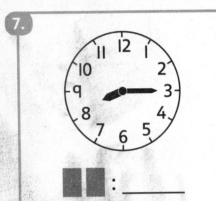

☐☐ : _____

**8.**

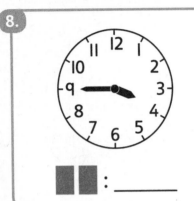

☐☐ : _____

**9.**

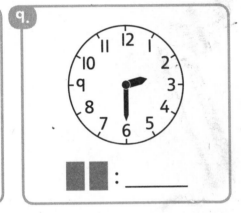

☐☐ : _____

**10.**

☐☐ : _____

**11.**

☐☐ : _____

**12.**

☐☐ : _____

**13.**

☐☐ : _____

**14.**

☐☐ : _____

**15.**

☐☐ : _____

138

Measurement and Data 2-30

☐ Where is the hour hand pointing?

**I.**

between ___7___

and ___8___

**2.**

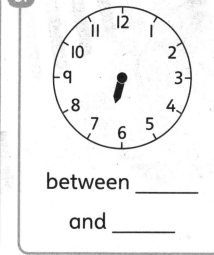

between _____

and _____

**3.**

between _____

and _____

**4.**

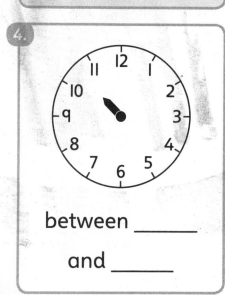

between _____

and _____

**5.**

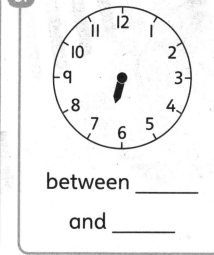

between _____

and _____

**6.**

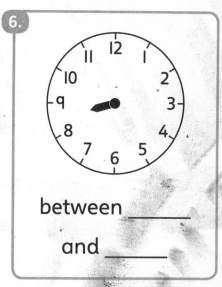

between _____

and _____

**7.**

between _____

and _____

**8.**

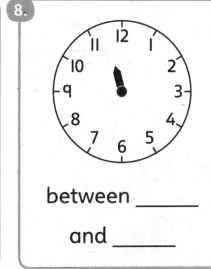

between _____

and _____

**q.**

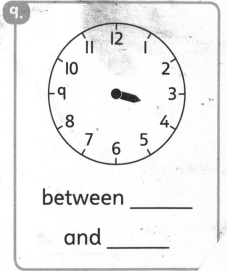

between _____

and _____

# What time is it?

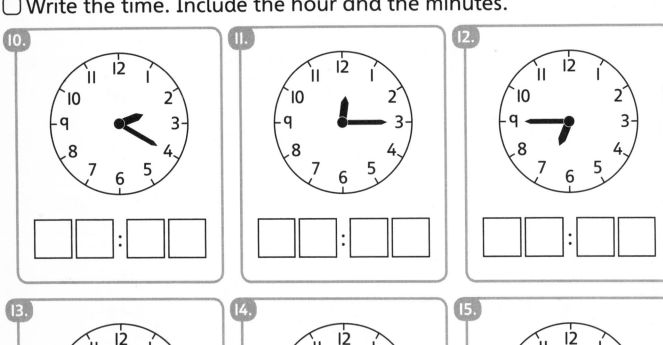

The hour hand is between 4 and 5. So we know the hour is 4.

The minute hand is pointing at the I. So we know it is 5 minutes after the hour.

| | 4 | : | 0 | 5 |

The **hour** goes here.  The **minutes** go here.

Write the time. Include the hour and the minutes.

**10.**

**11.**

**12.**

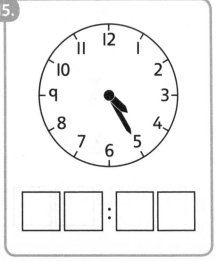

**13.**

**14.**

**15.**

## ☐ Write the time.

**16.**

☐☐ : ☐☐

**17.**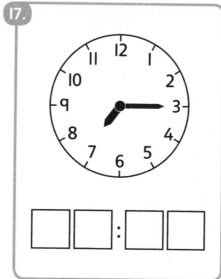

☐☐ : ☐☐

**18.**

☐☐ : ☐☐

**19.**

☐☐ : ☐☐

**20.**

☐☐ : ☐☐

**21.**

☐☐ : ☐☐

**22.**

☐☐ : ☐☐

**23.**

☐☐ : ☐☐

**24.**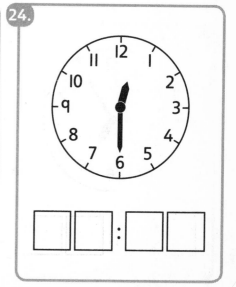

☐☐ : ☐☐

○ Write the time in words.

25.
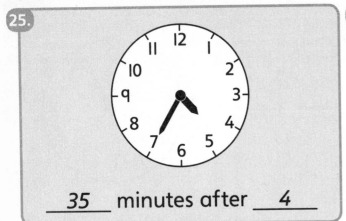
_35_ minutes after _4_

26.

_____ minutes after _____

27.

_____ minutes after _____

28.

_____ minutes after _____

29.
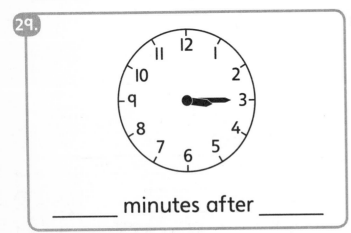
_____ minutes after _____

30.
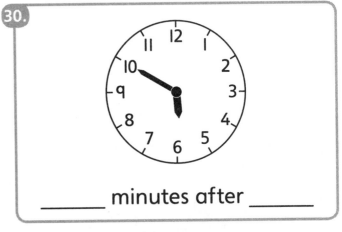
_____ minutes after _____

31.

The hour hand is between 6 and 7.
The minute hand points at the 3.
What time is it?

**Measurement and Data 2-31**

# MD2-32 Time to the 5 Minutes on a Digital Clock

 is the same time as

☐ Write the time in numbers and in words.

**1.**

____4____ : ____55____

*fifty-five minutes*

*after four*

**2.**

____ : ____

_____

**3.**

**11:05**

____ : ____

_____

**4.**

**5:15**

____ : ____

_____

**5.**

____ : ____

_____

**6.**

____ : ____

_____

Write the time the digital clock shows.
Then draw the time on the analog clock.

**7.**

___10___ : ___25___

**8.**

_____ : _____

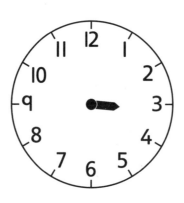

**9.**

_____ : _____

**10.**

_____ : _____

**11.**

_____ : _____

**12.**

_____ : _____

It is half an hour after 8:00 or 30 minutes after 8:00.

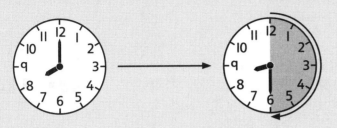

half past 8       8:30

☐ Write the time in two ways.

**1.**

half past _____

_____ : 30

**2.**

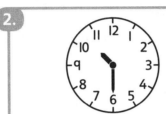

half past _____

_____ : 30

**3.**

half past _____

_____ : 30

**4.**

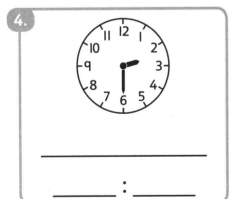

_____

_____ : _____

**5.**

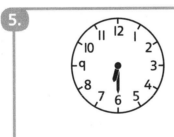

_____

_____ : _____

**6.**

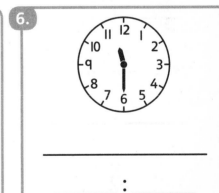

_____

_____ : _____

☐ Use a toy clock to show these times.
☐ Circle the two that are the same.

**7.**

| 12:30 | half past 3 | 4:30 |
|---|---|---|
| 5:30 | half past 7 | half past 12 |

# Look at where the hour hand is.
☐ Draw the minute hand at 12 or 6.
☐ Write the time.

**8.**

_half past 2_

**9.**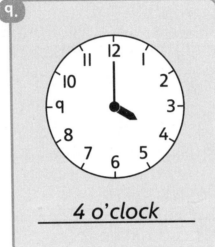

_4 o'clock_

**10.**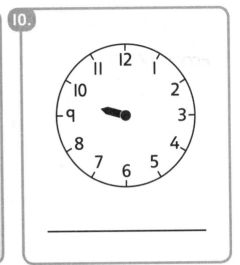

_____

**11.**

_____

**12.**

_____

**13.**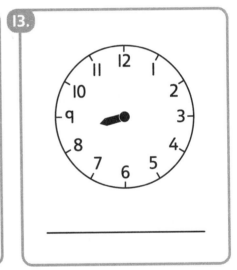

_____

**14.**

_____

**15.**

_____

**16.**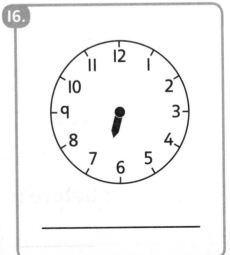

_____

**Measurement and Data 2-33**

From 12 o'clock midnight to 12 o'clock noon is called **a.m.**

Rob wakes up at 7 o'clock in the morning. We write **7 a.m.**

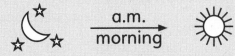

a.m. morning

From 12 o'clock noon to 12 o'clock midnight is called **p.m.**

Rob goes to bed at 8 o'clock in the evening. We write **8 p.m.**

p.m. afternoon and evening

☐ Is it **a.m.** or **p.m.**?

**1.**
7 o'clock in the morning

_____

**2.**
2 o'clock in the afternoon

_____

**3.**
9 o'clock in the evening

_____

**4.**
10 o'clock at night

_____

**5.**
4 o'clock in the afternoon

_____

**6.**
3 o'clock in the morning

_____

**7. BONUS**
2 hours **before** noon

_____

**8. BONUS**
3 hours **after** noon

_____

**9. BONUS**
1 hour **before** midnight

_____

**10. BONUS**
4 hours **after** midnight

_____

The timeline shows what Zara does every half hour
in the morning.

☐ Fill in the missing times.

**II.**

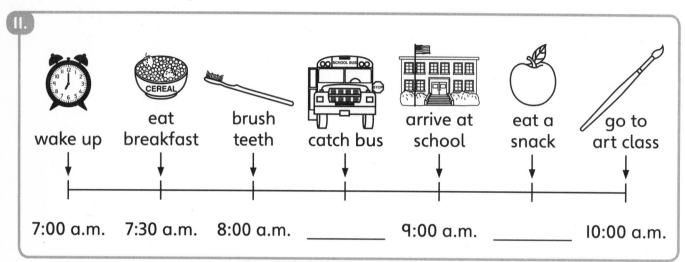

wake up    eat breakfast    brush teeth    catch bus    arrive at school    eat a snack    go to art class

7:00 a.m.    7:30 a.m.    8:00 a.m.    _____    9:00 a.m.    _____    10:00 a.m.

☐ Use Zara's timeline to answer the question.

**12.**

At what time does Zara eat breakfast? _____

**13.**

What time does Zara catch the bus? _____

**14.**

At what time does Zara go to art class? _____

**15.**

When does Zara arrive at school? _____

**16.**

What does Zara do at 7:00 a.m.?

_____

**17.**

What does Zara do at 8:00 a.m.?

_____

☐ Count by the hour.

**18.**

| Time Now | An Hour Later | An Hour Later | An Hour Later |
|---|---|---|---|
| 3:00 p.m. | 4:00 p.m. | 5:00 p.m. | |
| 7:00 a.m. | | | |
| 2:30 p.m. | 3:30 p.m. | 4:30 p.m. | |
| 8:30 a.m. | | | |

☐ Count by the half hour.

**19.**

| Time Now | A Half Hour Later | A Half Hour Later | A Half Hour Later |
|---|---|---|---|
| 2:00 a.m. | 2:30 a.m. | 3:00 a.m. | 3:30 a.m. |
| 5:00 p.m. | | | |
| 9:30 a.m. | | | |
| 7:30 p.m. | | | |

**20. BONUS**

Draw a timeline of your school day. Show what you do at different times of the day.

**21. BONUS**

Jenny goes to baseball practice 3 hours after school ends. School ends at 4:00 p.m. What time does Jenny go to practice?

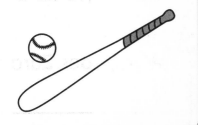

# MD2-35 Skip Counting by Different Numbers

Skip count by ⟨5s⟩, then by ⟨1s⟩.

**1.**

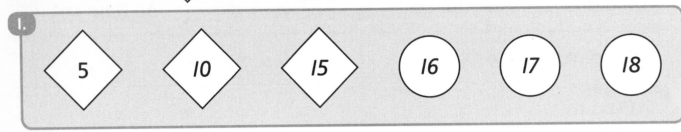

**2.**

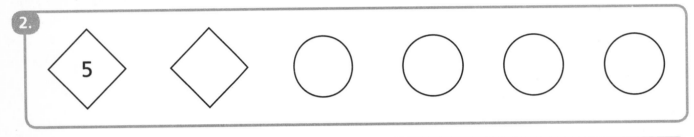

**3.**

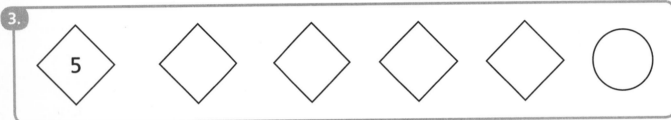

Skip count by |10s|, then by ⟨1s⟩.

**4.**

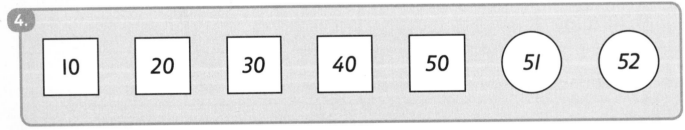

**5.**

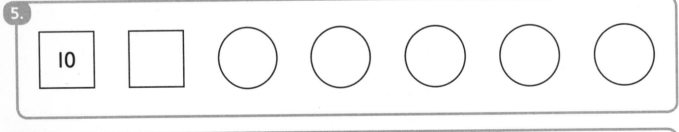

**6.**

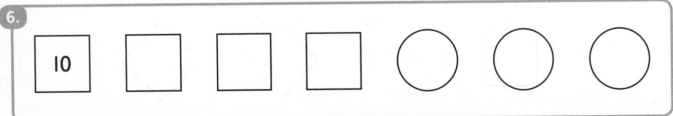

☐ Skip count by ⬜10s⬜, then by ◇5s◇.

**7.**
| 10 | 20 | 30 | 40 | ◇45◇ | ◇50◇ |

**8.**
| 10 | ⬜ | ⬜ | ◇ | ◇ | ◇ |

**9.**
| 10 | ⬜ | ⬜ | ⬜ | ⬜ | ◇ |

☐ Skip count by ⬜10s⬜, then by ◇5s◇, then by ◯1s◯.

**10.**
| 10 | ⬜ | ⬜ | ◇ | ◯ | ◯ |

**11.**
| 10 | ◇ | ◇ | ◯ | ◯ | ◯ |

**12.**
| 10 | ⬜ | ◇ | ◇ | ◇ | ◯ |

☐ Skip count by ⬡25s⬡, then by ▢10s▢.

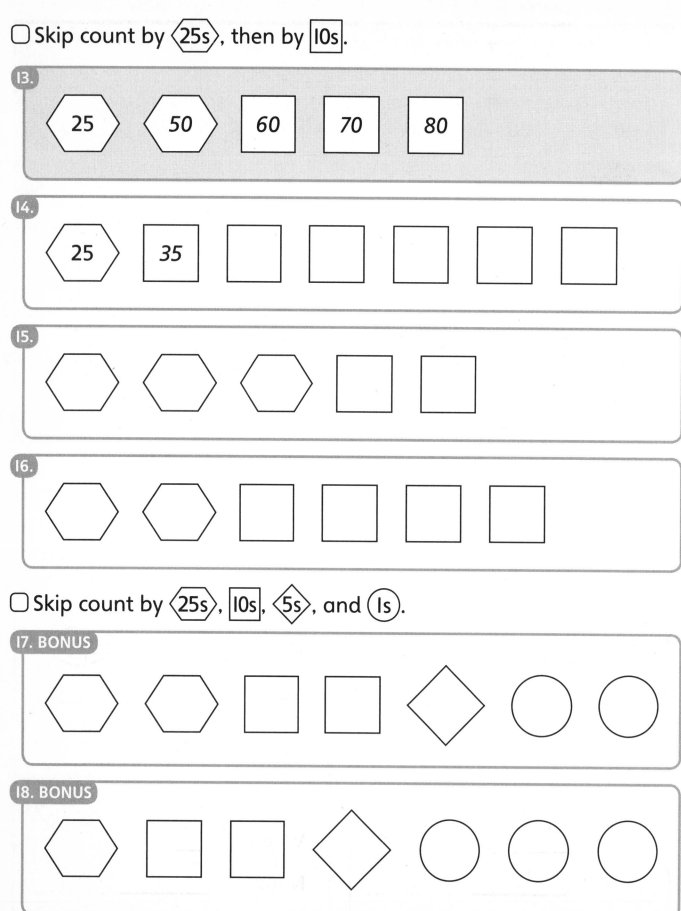

**13.**

⬡ 25 ⬡   ⬡ 50 ⬡   ▢ 60 ▢   ▢ 70 ▢   ▢ 80 ▢

**14.**

⬡ 25 ⬡   ▢ 35 ▢   ▢   ▢   ▢   ▢   ▢

**15.**

⬡   ⬡   ⬡   ▢   ▢

**16.**

⬡   ⬡   ▢   ▢   ▢   ▢

☐ Skip count by ⬡25s⬡, ▢10s▢, ◇5s◇, and ◯1s◯.

**17. BONUS**

⬡   ⬡   ▢   ▢   ◇   ◯   ◯

**18. BONUS**

⬡   ▢   ▢   ◇   ◯   ◯   ◯

# MD2-36 Coin Values

**penny**     **nickel**     **dime**     **quarter**

1¢        5¢        10¢        25¢

You write
¢ for cents.

☐ Match the coin with the name.

1.

dime      quarter      penny      nickel

☐ Write the value of the coin.
☐ Write the name of the coin.

2.

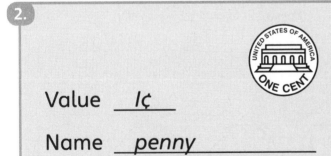

Value ___1¢___

Name ___penny___

3.

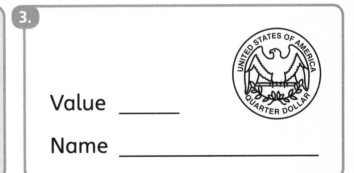

Value _____

Name _____

4.

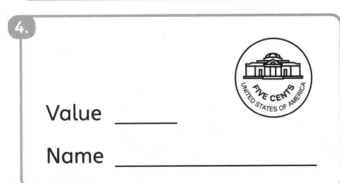

Value _____

Name _____

5.

Value _____

Name _____

☐ Answer the questions.

**6.**

How many pennies make a nickel? _____

How many pennies make a dime? _____

How many pennies make a quarter? _____

**7.**

How many nickels make a dime? _____

How many nickels make a quarter? _____

☐ Skip count by coin value.

**8.**

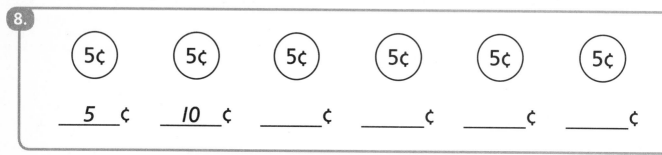

  __5__ ¢    __10__ ¢    _____ ¢    _____ ¢    _____ ¢    _____ ¢

**9.**

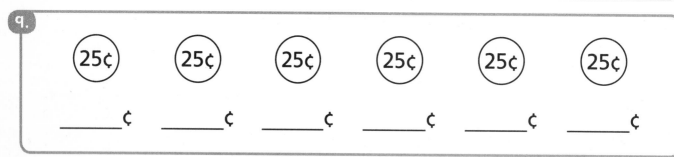

_____ ¢    _____ ¢    _____ ¢    _____ ¢    _____ ¢    _____ ¢

☐ Circle the greater amount.

**10.**

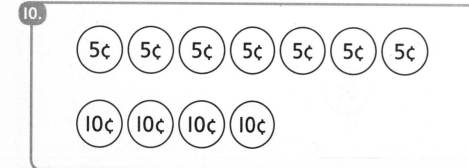

**Measurement and Data 2-36**

Do Alice and Bo have the same amount of money?
Write **Yes** or **No**.

| 11. Alice's Amount | Bo's Amount | | | | | Same Amount? |
|---|---|---|---|---|---|---|
| 10¢ <br> _10_ ¢ | 5¢ <br> _5_ ¢ | 5¢ <br> _10_ ¢ | 5¢ <br> _15_ ¢ | | | No |
| 25¢ <br> ___ ¢ | 10¢ <br> ___ ¢ | 10¢ <br> ___ ¢ | 10¢ <br> ___ ¢ | | | |
| 5¢ <br> ___ ¢ | 1¢ <br> ___ ¢ | 1¢ <br> ___ ¢ | 1¢ <br> ___ ¢ | 1¢ <br> ___ ¢ | 1¢ <br> ___ ¢ | |
| 25¢ <br> ___ ¢ | 5¢ <br> ___ ¢ | 5¢ <br> ___ ¢ | 5¢ <br> ___ ¢ | 5¢ <br> ___ ¢ | | |
| 25¢ <br> ___ ¢ | 5¢ <br> ___ ¢ | 5¢ <br> ___ ¢ | 5¢ <br> ___ ¢ | 5¢ <br> ___ ¢ | 5¢ <br> ___ ¢ | |
| **BONUS:** <br> 25¢ <br> ___ ¢ | 10¢ <br> ___ ¢ | 10¢ <br> ___ ¢ | 5¢ <br> ___ ¢ | | | |

# MD2-37 Counting Coins

☐ Count the money by coin value.

1.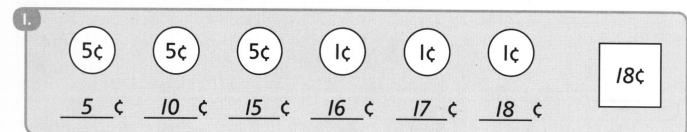

    5 ¢    10 ¢    15 ¢    16 ¢    17 ¢    18 ¢     18¢

2.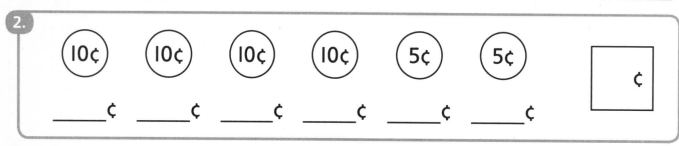

    ___ ¢    ___ ¢    ___ ¢    ___ ¢    ___ ¢    ___ ¢     ¢

3.

    ___ ¢    ___ ¢    ___ ¢    ___ ¢    ___ ¢     ¢

4.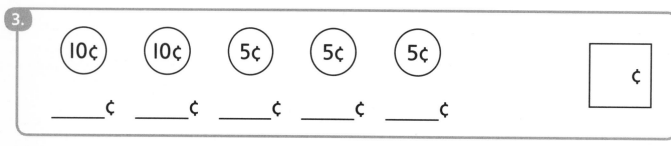

    ___ ¢    ___ ¢    ___ ¢    ___ ¢    ___ ¢    ___ ¢     ¢

5.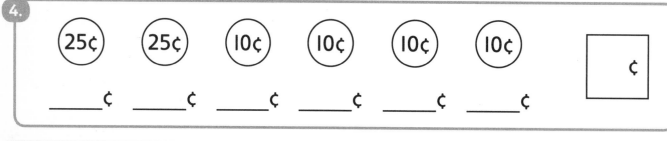

    ___ ¢    ___ ¢    ___ ¢    ___ ¢    ___ ¢    ___ ¢     ¢

6.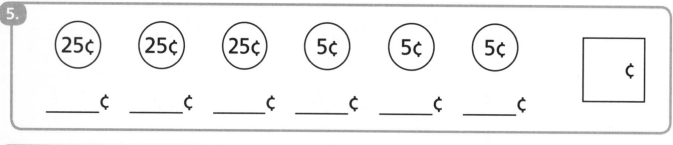

    ___ ¢    ___ ¢    ___ ¢    ___ ¢    ___ ¢    ___ ¢     ¢

     **Measurement and Data 2-37**

☐ Count the money by coin value.

**7.**

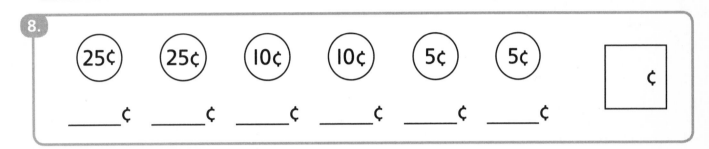

10¢  10¢  5¢  5¢  5¢  1¢    [ ____ ¢ ]

___¢  ___¢  ___¢  ___¢  ___¢  ___¢

**8.**

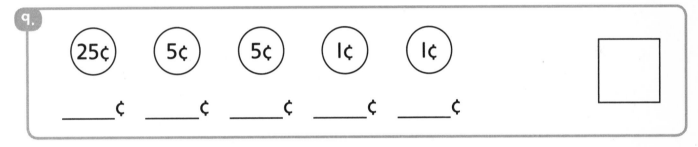

25¢  25¢  10¢  10¢  5¢  5¢    [ ____ ¢ ]

___¢  ___¢  ___¢  ___¢  ___¢  ___¢

**9.**

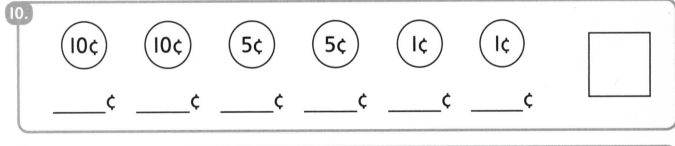

25¢  5¢  5¢  1¢  1¢    [ ____ ]

___¢  ___¢  ___¢  ___¢  ___¢

**10.**

10¢  10¢  5¢  5¢  1¢  1¢    [ ____ ]

___¢  ___¢  ___¢  ___¢  ___¢  ___¢

**11.**

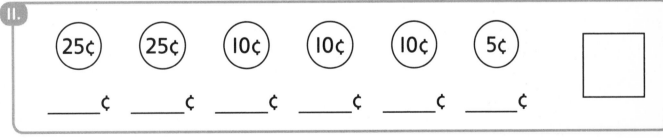

25¢  25¢  10¢  10¢  10¢  5¢    [ ____ ]

___¢  ___¢  ___¢  ___¢  ___¢  ___¢

**12.**

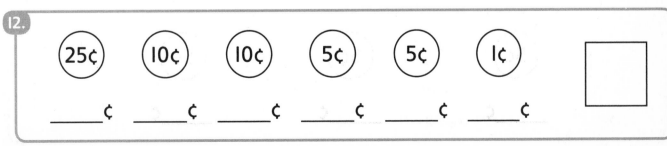

25¢  10¢  10¢  5¢  5¢  1¢    [ ____ ]

___¢  ___¢  ___¢  ___¢  ___¢  ___¢

☐ Write the coin values from largest to smallest.
☐ Count the money by coin value.

13.
25¢  5¢  1¢      25¢ ◯ ◯      ☐

14.
10¢  5¢  5¢      ◯ ◯ ◯      ☐

15.
10¢  25¢  25¢  5¢      ◯ ◯ ◯ ◯      ☐

16.
25¢  1¢  10¢  5¢      ◯ ◯ ◯ ◯      ☐

17.
Jake has 3 quarters and 2 dimes.
How many cents does he have?
25¢  25¢  25¢  10¢  10¢      ☐

18. BONUS
Sally has 5 dimes, 2 nickels, and 3 pennies.
How many cents does she have?

# MD2-38 More Counting Coins

☐ Circle groups of 10¢ using blue.
☐ Count the money.

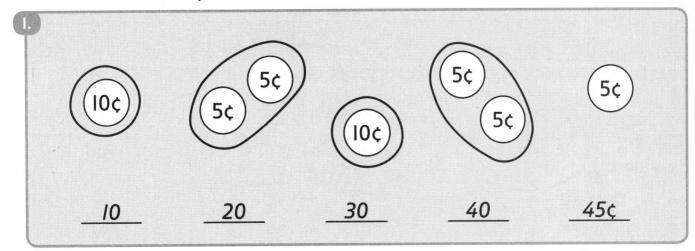

1.

| 10 | 20 | 30 | 40 | 45¢ |

2.

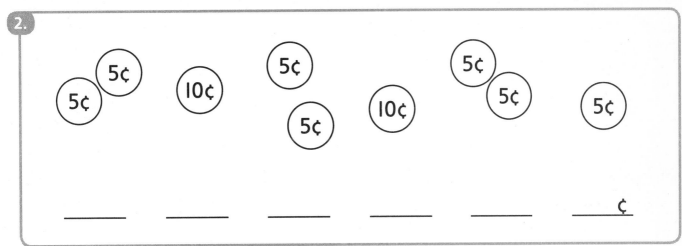

___   ___   ___   ___   ___   ___ ¢

☐ Circle groups of 25¢ using red.
☐ Count the money.

3.

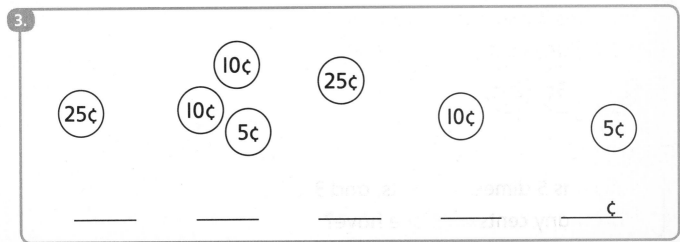

___   ___   ___   ___   ___ ¢

☐ Circle all groups of 25¢ using red.
☐ Circle all groups of 10¢ using blue.
☐ Count the money.

**4.**

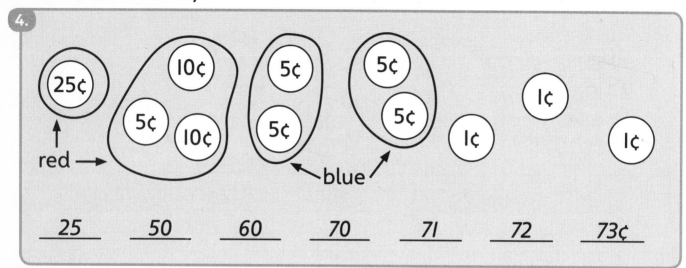

25    50    60    70    71    72    73¢

**5.**

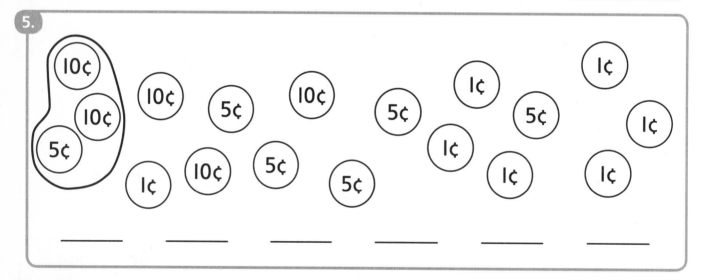

_____    _____    _____    _____    _____

**6. BONUS**

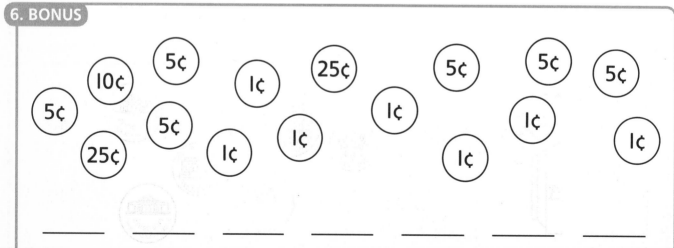

_____    _____    _____    _____    _____

**Measurement and Data 2-38**

# Does Liz have enough money?

⬜ Circle **Yes** or **No**.

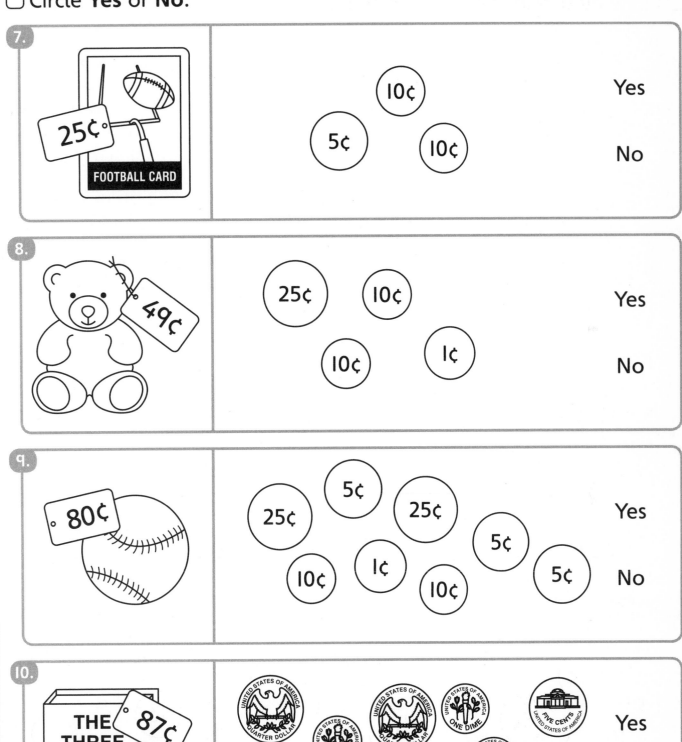

**7.** 25¢ FOOTBALL CARD — 10¢ 5¢ 10¢ — Yes / No

**8.** 49¢ teddy bear — 25¢ 10¢ 10¢ 1¢ — Yes / No

**9.** 80¢ baseball — 25¢ 5¢ 25¢ 10¢ 1¢ 10¢ 5¢ 5¢ — Yes / No

**10.** 87¢ THE THREE BEARS — Quarter Dollar, One Dime, Quarter Dollar, One Dime, One Dime, Five Cents, Five Cents, One Cent, Five Cents, Five Cents — Yes / No

☐ Draw coins to make 12¢.

**11.**

Use 3 coins.

**12.**

Use 4 coins.

☐ Draw coins to make 27¢.

**13.**

Use 3 coins.

**14.**

Use 5 coins.

**15.**

Use 6 coins.

**16.**

Use 7 coins.

**17.** Draw coins to make each amount using the fewest coins.

8¢       15¢       21¢       31¢       36¢

**18.** Show how many different ways you can make 13¢ using nickels and pennies.

# MD2-39 The Dollar

☐ Count on by 1¢.

**1.**

97¢, __98__ , _____, _____, _____, _____, _____ ¢

☐ Skip count by 5¢.

**2.**

90¢, __95__ , __100__ , __105__ ¢

**3.**

95¢, _____, _____, _____ ¢

**4.**

80¢, _____, _____, _____, _____, _____, _____ ¢

☐ Skip count by 10¢.

**5.**

80¢, __90__ , _____, _____ ¢

**6.**

95¢, __105__ , _____, _____ ¢

**7.**

70¢, _____, _____, _____, _____, _____, _____ ¢

**8.**

75¢, __85__ , _____, _____, _____, _____, _____ ¢

☐ Skip count by 25¢.

**9.**

25¢, __50__ , _____, _____, _____, _____, _____, _____ ¢

A **dollar** has the same value
as 100¢ or 100 pennies.

☐ Count the money by coin value.
☐ Is the value more or less than a dollar? Circle the answer.

**10.**

(25¢) (25¢) (25¢) (25¢) (25¢)

_____¢  _____¢  _____¢  _____¢  _____¢

more

less

**11.**

(25¢) (25¢) (10¢) (10¢) (10¢)

_____¢  _____¢  _____¢  _____¢  _____¢

more

less

**12.**

(25¢) (25¢) (25¢) (10¢) (10¢) (10¢)

_____¢  _____¢  _____¢  _____¢  _____¢  _____¢

more

less

**13.**

(25¢) (25¢) (25¢) (10¢) (5¢) (5¢)

_____¢  _____¢  _____¢  _____¢  _____¢  _____¢

more

less

**14.**

(25¢) (25¢) (25¢) (10¢) (10¢) (1¢)

_____¢  _____¢  _____¢  _____¢  _____¢  _____¢

more

less

**Measurement and Data 2-39**

☐ Shade coins to make a dollar.

**15.**
(25¢) (25¢) (10¢)
(25¢) (5¢) (25¢)

**16.**
(25¢) (25¢) (10¢)
(25¢) (5¢) (10¢)

**17.**
(25¢) (10¢) (10¢) (25¢)
(10¢) (5¢) (10¢) (10¢)

**18.**
(25¢) (10¢) (10¢) (5¢)
(10¢) (10¢) (10¢) (25¢)

**19.**
(25¢) (10¢) (25¢) (10¢)
(10¢) (5¢) (10¢) (5¢)

**20.**
(25¢) (10¢) (25¢) (5¢)
(10¢) (25¢) (10¢) (10¢)

☐ Shade enough coins in the bottom row to make a dollar.

**21.**
(25¢) (25¢) (25¢)

(25¢) (25¢) (10¢) (5¢) (5¢)

**22.**
(25¢) (25¢) (10¢) (10¢)

(25¢) (25¢) (10¢) (5¢)

**23.** Draw the number of dimes that make a dollar.

**24.** Draw the number of nickels that make a dollar.

**Measurement and Data 2-39**                               165

# MD2-40 Dollar Notation

You can write 143¢ in two ways.

| cent notation | dollar notation |
|---|---|
| 143¢ | $1.43 |

dollars    dimes    pennies

You write $1.00 for one dollar.

☐ Complete the table.

**1.**

| Cent Notation | Dollars | Dimes | Pennies | Dollar Notation |
|---|---|---|---|---|
| 152¢ | 1 | 5 | 2 | $1.52 |
| 219¢ | | | | |
| 425¢ | | | | |
| 554¢ | | | | |
| 816¢ | | | | |
| 205¢ | | | | |
| 300¢ | | | | |
| 265¢ | | | | |

☐ **BONUS:** Write the amounts in cent notation.

**2.**

$1.25  _____

$4.07  _____

$0.26  _____

**3.**

$4.00  _____

$1.76  _____

$2.01  _____

Complete the table.

| Dollar Amount | Cent Amount | Total |
|---|---|---|
| $1  $1 <br> = _2 dollars_ | 25¢ 10¢ <br> = _35¢_ | _$2.35_ |
| $1  $1  $1 <br> = _____ | 5¢ 5¢ 1¢ <br> = _____ | _____ |
| $1  $1 <br> = _____ | 10¢ 10¢ 10¢ <br> = _____ | _____ |
| $1 <br> = _____ | 25¢ 25¢ 5¢ <br> = _____ | _____ |
| $1  $1  $1 <br> = _____ | 5¢ 1¢ 1¢ <br> = _____ | _____ |
| $1  $1 <br> = _____ | 5¢ 5¢ 1¢ 1¢ <br> = _____ | _____ |
| $1  $1  $1 <br> = _____ | 25¢ 25¢ 25¢ <br> = _____ | _____ |

☐ Count the money by coin value.
☐ Write the total amount in dollar notation.

**5.**

| Coins | Dollar Notation |
|---|---|
| 25¢ 25¢ 25¢ 25¢ 10¢ | $1.10 |
| 25¢ 25¢ 25¢ 25¢ 1¢ 1¢ | _____ |
| 25¢ 25¢ 25¢ 25¢ 25¢ | _____ |
| 25¢ 25¢ 25¢ 25¢ 1¢ | _____ |
| 25¢ 25¢ 25¢ 25¢ 10¢ 5¢ | _____ |
| 25¢ 25¢ 25¢ 25¢ 10¢ 1¢ | _____ |
| 25¢ 25¢ 25¢ 25¢ 5¢ 5¢ | _____ |
| 25¢ 25¢ 25¢ 25¢ 5¢ | _____ |
| **BONUS:** | |
| 25¢ 25¢ 25¢ 10¢ 10¢ 10¢ | _____ |
| 25¢ 25¢ 25¢ 10¢ 10¢ 10¢ 10¢ | |

# MD2-4I Adding Money

Micky adds coins to her bag.
☐ How much money does she have now?

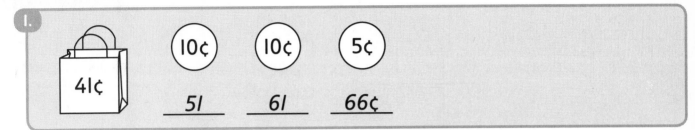

1. 41¢   10¢   10¢   5¢
         51    61    66¢

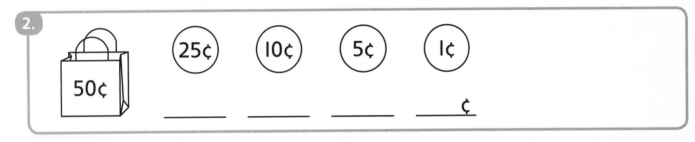

2. 50¢   25¢   10¢   5¢   1¢
         ___   ___   ___   ___ ¢

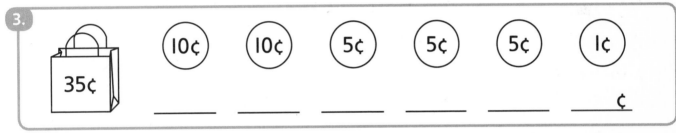

3. 35¢   10¢   10¢   5¢   5¢   5¢   1¢
         ___   ___   ___   ___   ___   ___ ¢

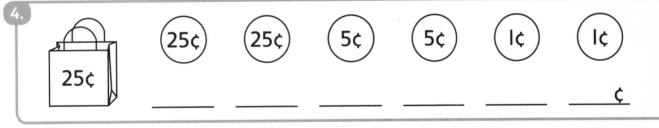

4. 25¢   25¢   25¢   5¢   5¢   1¢   1¢
         ___   ___   ___   ___   ___   ___ ¢

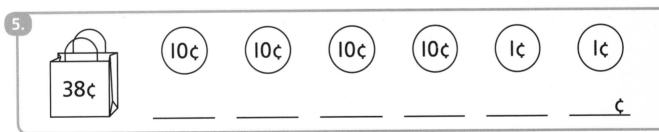

5. 38¢   10¢   10¢   10¢   10¢   1¢   1¢
         ___   ___   ___    ___   ___   ___ ¢

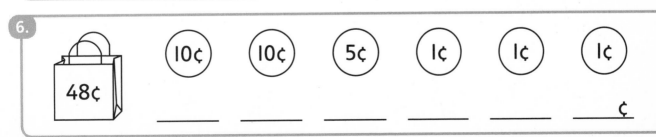

6. 48¢   10¢   10¢   5¢   1¢   1¢   1¢
         ___   ___   ___   ___   ___   ___ ¢

Andy adds coins to his bag.

☐ How much money does he have now?

**7.**

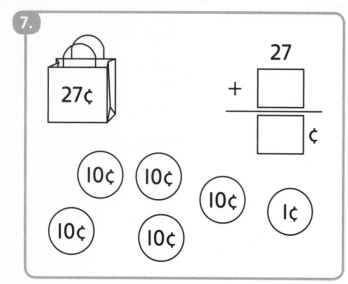

**8.**

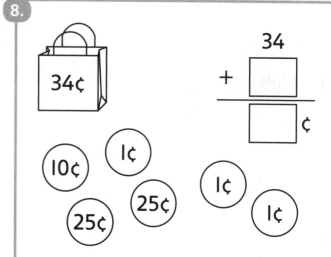

**9.**

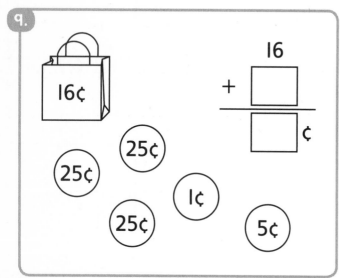

**10.**

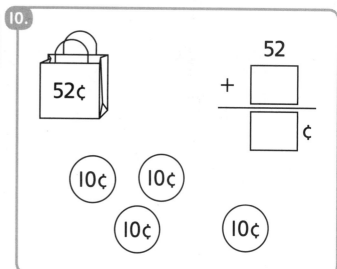

**11.**

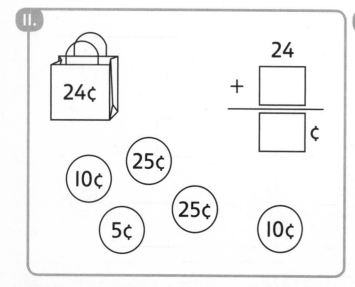

**12. BONUS**

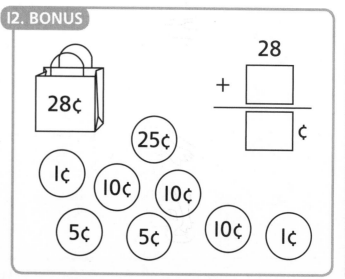

**Measurement and Data 2-41**

# MD2-42 Subtracting Money

Marta pays for a sticker.

☐ How much money will she get back?

**1.**

25¢

| 25 | – | 18 | = | 7 |

*She will get back 7¢*

_____

**2.**

71¢   25¢   25¢   25¢

☐☐ – ☐ = ☐

_____

_____

**3.**

55¢   25¢   25¢   10¢

☐☐ – ☐ = ☐

_____

_____

**4. BONUS**

94¢   $1

| 100 | – | ☐ | = | ☐ |

_____

_____

Ravi pays for a toy.
⬜ How much money will he get back?

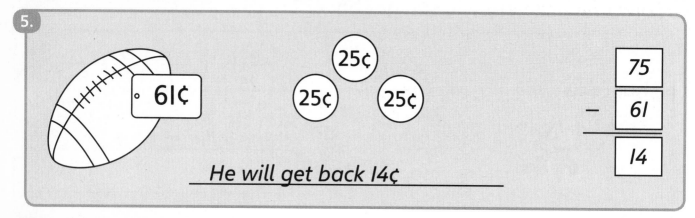

**5.**

He will get back 14¢

|  | 75 |
|---|---|
| − | 61 |
|  | 14 |

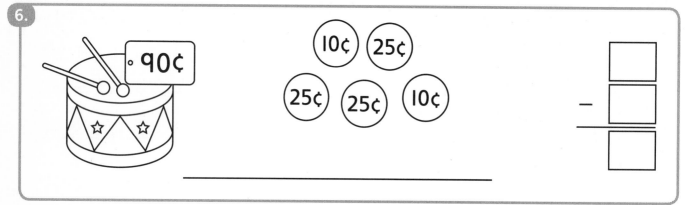

**6.**

_____

|  | ☐ |
|---|---|
| − | ☐ |
|  | ☐ |

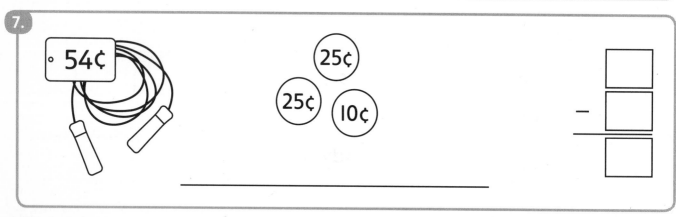

**7.**

_____

|  | ☐ |
|---|---|
| − | ☐ |
|  | ☐ |

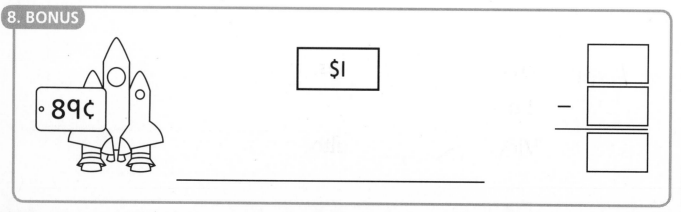

**8. BONUS**

$1

_____

|  | ☐ |
|---|---|
| − | ☐ |
|  | ☐ |

# MD2-43 Money Word Problems

☐ Write an addition or subtraction sentence.
☐ Solve the problem.

**1.**

Rani has 17¢.

She found 3 dimes.

Now Rani has _____¢.

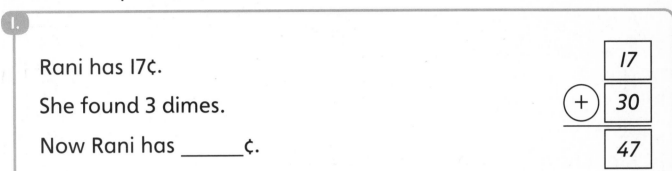

**2.**

Mike has 60¢.

He gave his sister a quarter.

Mike has _____¢ left.

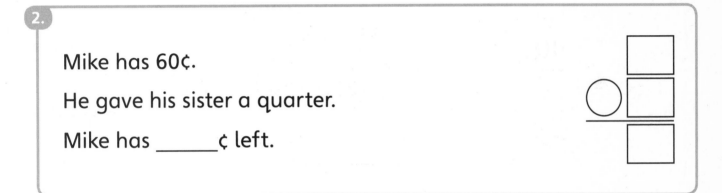

**3.**

Eddy has 2 dimes and 3 nickels.

Ray has 2 quarters.

Ray has _____¢ more than Eddy.

**4.**

Amit has 3 nickels and 7 pennies.

Vicky has 2 dimes.

Amit and Vicky have _____¢ altogether.

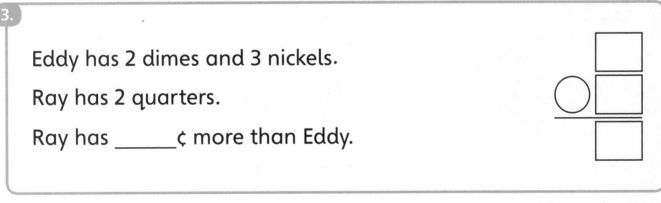

○ Write an addition or subtraction sentence.
○ Find the answer.

**5.**
Mary has 75¢ for a snack.
The snack costs 65¢.
How much change will she get?

**6.**
Kim has these coins.
How much more money does
she need to make $1.00?

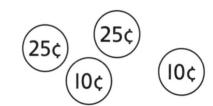

**7.**
Alex has 5 nickels, 5 pennies, 6 dimes, and 1 quarter.
Does he have more than a dollar?
If so, how much more?

**8.**
Nancy wants to buy a toy train for 99¢.
She has 2 quarters and 5 dimes.
Does she have enough money? Show how you know.

**9.**
Jay has 4 nickels and 3 dimes.
Fred has 3 nickels and 4 dimes.
Who has more money? How much more?

**10. BONUS**
Paul has 80¢. He needs a dollar to buy a comic book.
How much more money does he need?

# G2-1 Lines

| straight lines | curved lines | curved side |
|---|---|---|
|  | 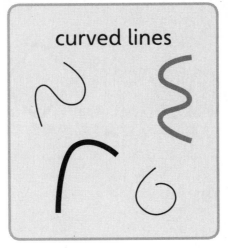 | 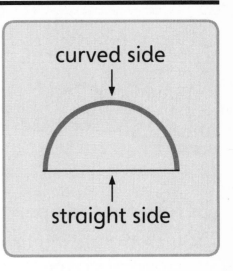 |
| | | straight side |

☐ Draw ✕ on all the shapes that have a **straight** side.

☐ Draw ◯ on all the shapes that have a **curved** side.

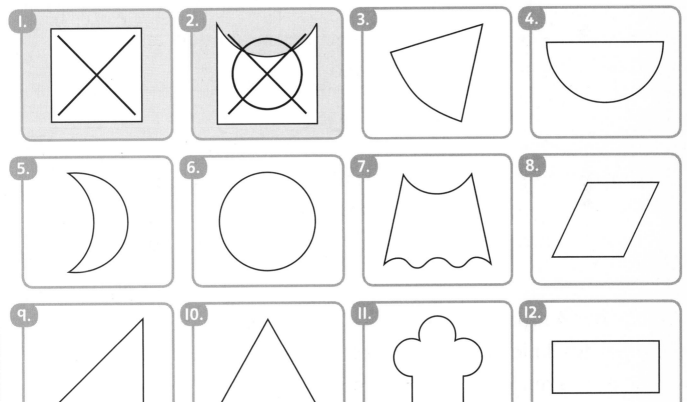

☐ Color the boxes of the shapes with ✕ and ◯.

**13.**

What letter do you see? _____

| open | closed |
|------|--------|
| 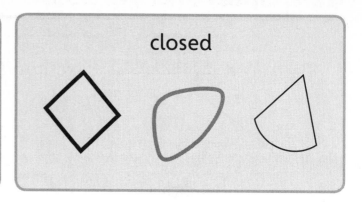 | |

☐ Cross out the open lines.
☐ Circle the closed lines.

**14.**

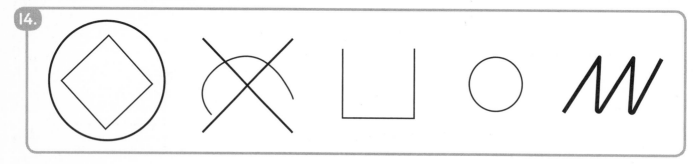

**15.**

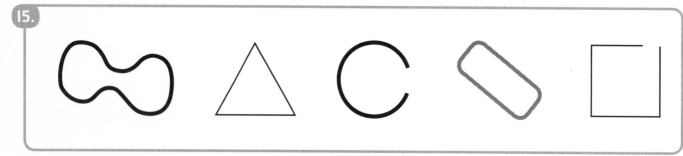

**16.**

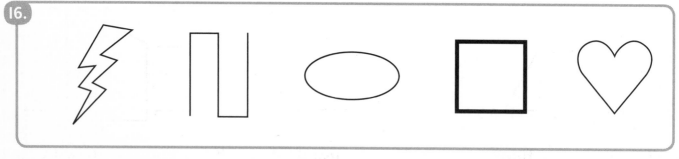

**17.**

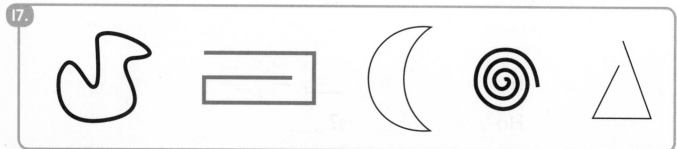

**Geometry 2-1**

# G2-2 Sides and Vertices

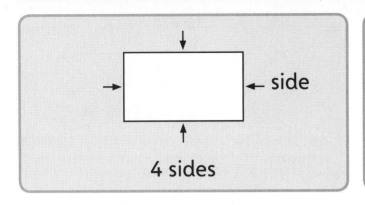 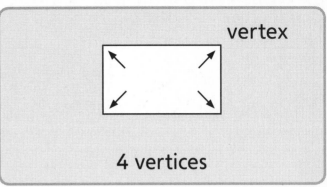

4 sides        4 vertices

☐ Count the sides.

**1.**

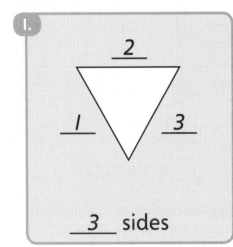

__3__ sides

**2.**

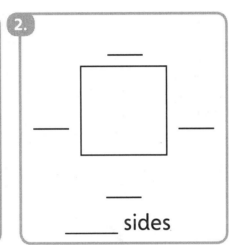

_____ sides

**3.**

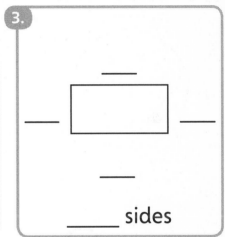

_____ sides

☐ Count the vertices.

**4.**

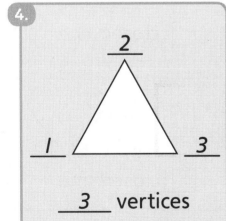

__3__ vertices

**5.**

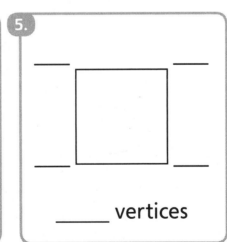

_____ vertices

**6.**

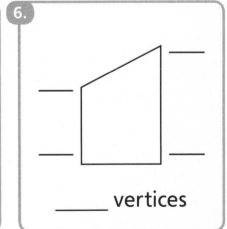

_____ vertices

**7.**

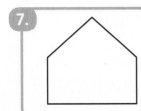

How many sides? _____

How many vertices? _____

○ Write ✓ if true. Write ✗ if not true.

**8.**
☑ 4 sides
☑ 4 vertices

**9.**
☐ 4 sides
☐ 4 vertices

**10.**
☐ 4 sides
☐ 4 vertices

**11.**
☐ 4 sides
☐ 4 vertices

**12.**
☐ 4 sides
☐ 4 vertices

**13.**
☐ 4 sides
☐ 4 vertices

**14.**
☐ 3 sides
☐ 3 vertices
☐ closed line

**15.**
☐ 3 sides
☐ 3 vertices
☐ closed line

**16.**
☐ 4 sides
☐ 4 vertices
☐ closed line

**17.**
☐ 4 sides
☐ 4 vertices
☐ closed line

**18.**
☐ 3 sides
☐ 3 vertices
☐ 3 straight sides

**19.**
☐ 4 sides
☐ 4 vertices
☐ 4 straight sides

**Geometry 2-2**

# G2-3 Squares and Rectangles

| squares | not squares |
|---|---|
| 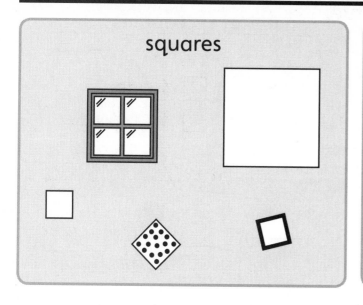 | 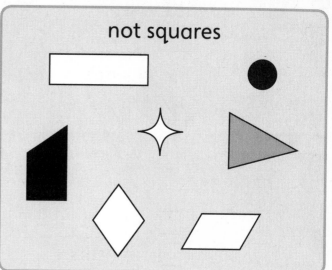 |

☐ Draw ✕ on the shapes that are **not** squares.

1.
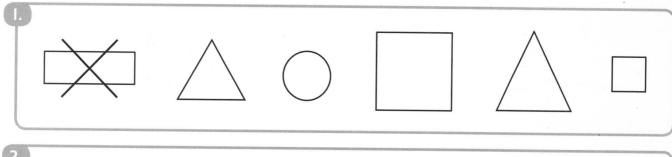

2.
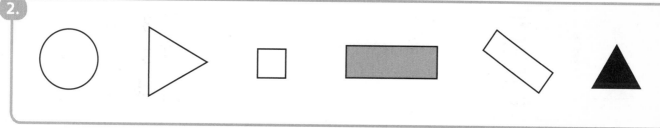

3.
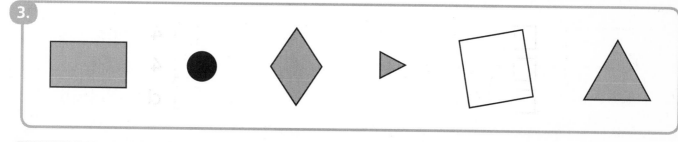

4.
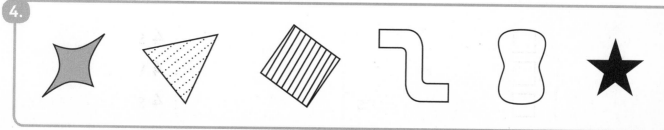

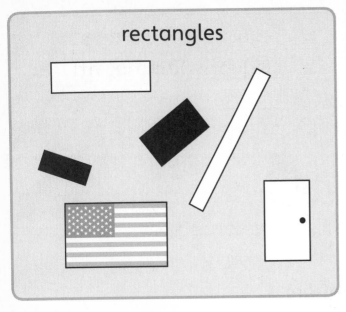

rectangles

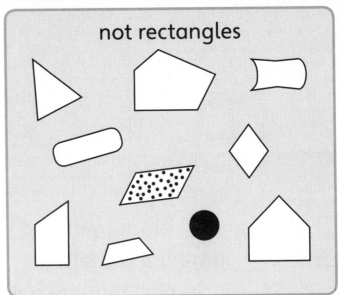

not rectangles

☐ Circle the rectangles.

5.

6.

7.

8.

| square corners | not square corners (square does not fit) |
|---|---|
| 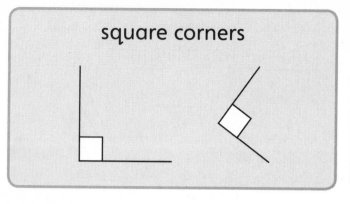 | 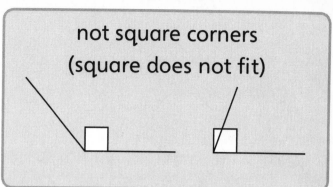 |

☐ Use a pattern block square to find square corners.

☐ Draw ✓ in square corners.

☐ Draw ✗ in corners that are not square.

| 9. | 10. | 11. |
|---|---|---|
| 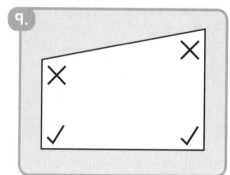 |  | 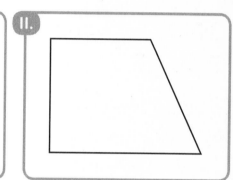 |

☐ Draw a shape with a square corner.

☐ Draw a shape with no square corners.

| 12. | 13. |
|---|---|
| square corner | no square corners |
| 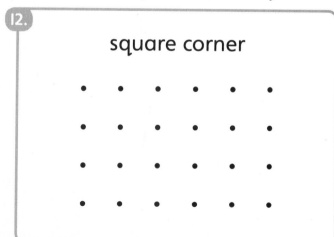 | 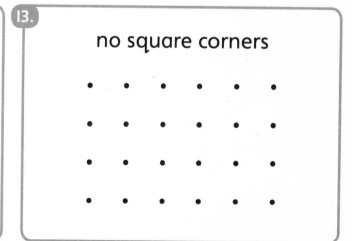 |

14.

This is not a rectangle. How do you know?

# G2-4 Polygons

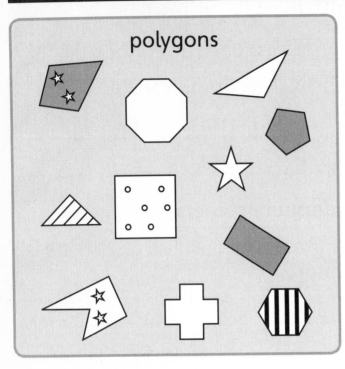

polygons

not polygons

☐ Circle the polygons.

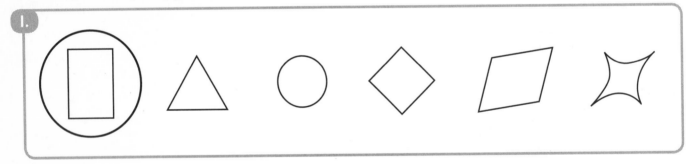

1.

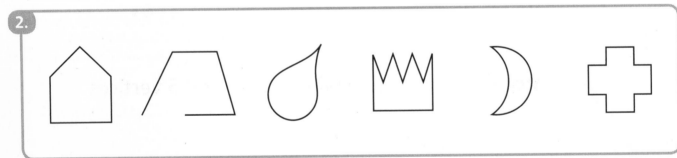

2.

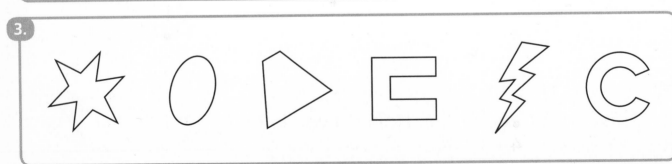

3.

**Geometry 2-4**

☐ Use a ruler. Connect the dots in order.
☐ Join the first and last dots.

**4.** A **triangle** is a polygon with 3 sides and 3 vertices.

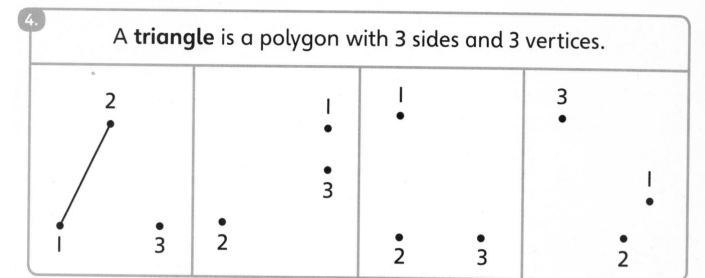

**5.** A **quadrilateral** is a polygon with 4 sides and 4 vertices.

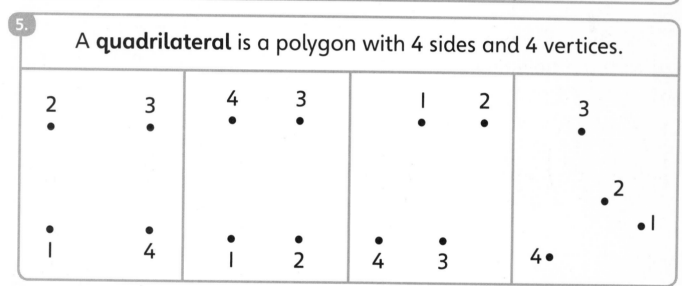

**6.** A **pentagon** is a polygon with 5 sides and 5 vertices.

☐ Use a ruler. Connect the dots in order.

☐ Join the first and last dots.

**7.**

A **hexagon** is a polygon with 6 sides and 6 vertices.

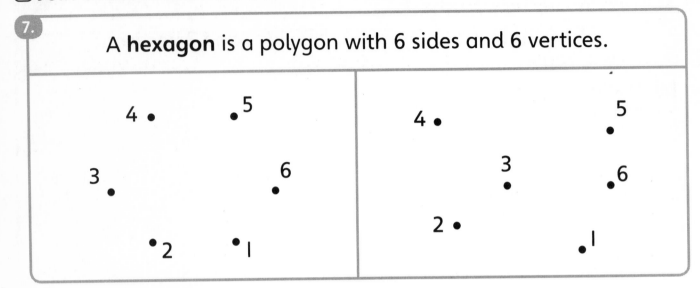

☐ Join the dots. Then name the shape.

**8.**

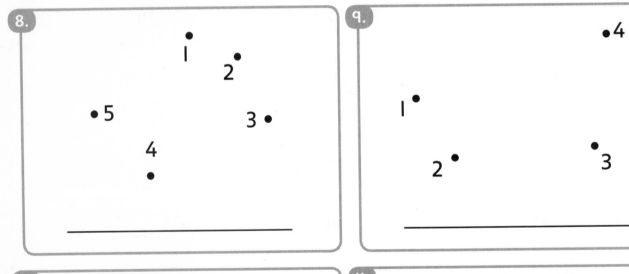

**9.**

**10.**

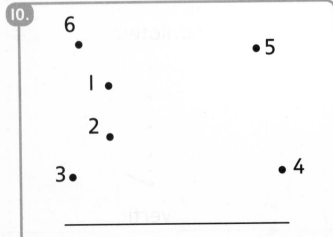

**11.**

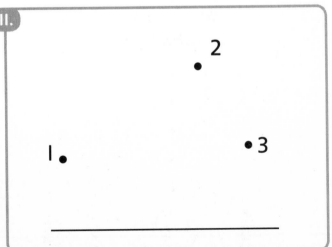

# G2-5 About Polygons

☐ Draw the missing sides to complete the shape.
☐ Count the vertices to check.

**1.**

rectangle

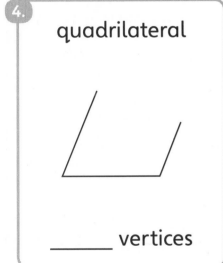

| | |
|---|---|
| 1 | 3 |
| 2 | 4 |

___4___ vertices

**2.**

square

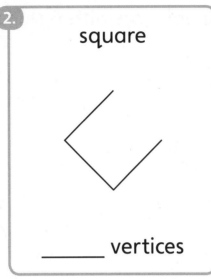

_____ vertices

**3.**

triangle

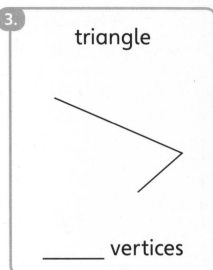

_____ vertices

**4.**

quadrilateral

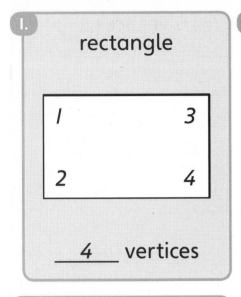

_____ vertices

**5.**

pentagon

_____ vertices

**6.**

hexagon

_____ vertices

**7.**

hexagon

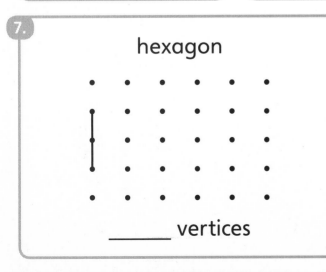

_____ vertices

**8.**

quadrilateral

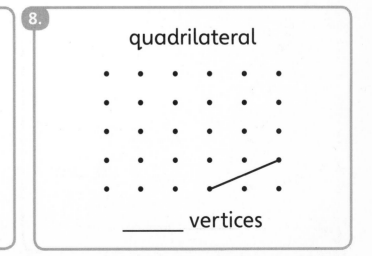

_____ vertices

hexagon                    pentagon              quadrilateral
rectangle                   square                  triangle

☐ How many sides? How many vertices?
☐ Name the shape. Use the words in the box.

**9.**
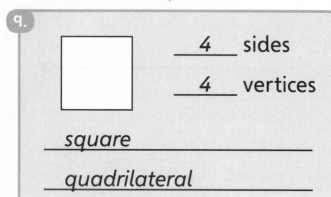
___4___ sides
___4___ vertices

_square_

_quadrilateral_

**10.**
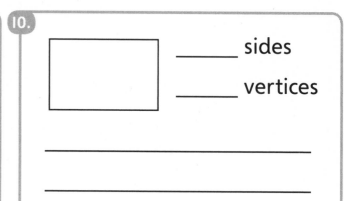
_____ sides
_____ vertices

_____

_____

**11.**
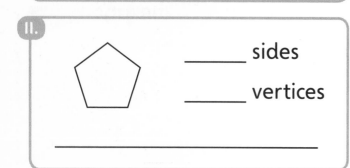
_____ sides
_____ vertices

_____

**12.**
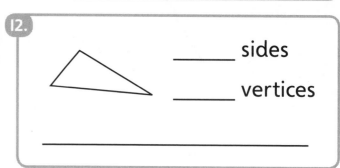
_____ sides
_____ vertices

_____

**13.**
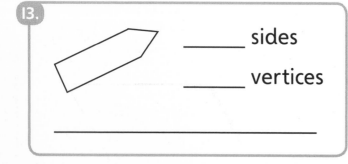
_____ sides
_____ vertices

_____

**14.**
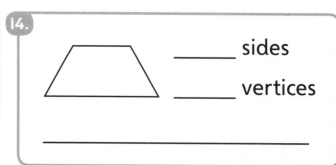
_____ sides
_____ vertices

_____

**15.**
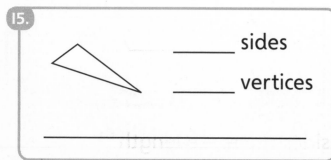
_____ sides
_____ vertices

_____

**16.**
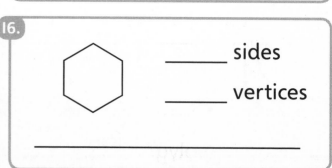
_____ sides
_____ vertices

_____

**Geometry 2-5**

# G2-6 Equal Sides

☐ Use a centimeter ruler to measure the sides.

☐ Color the sides that are the same length as the thick side.

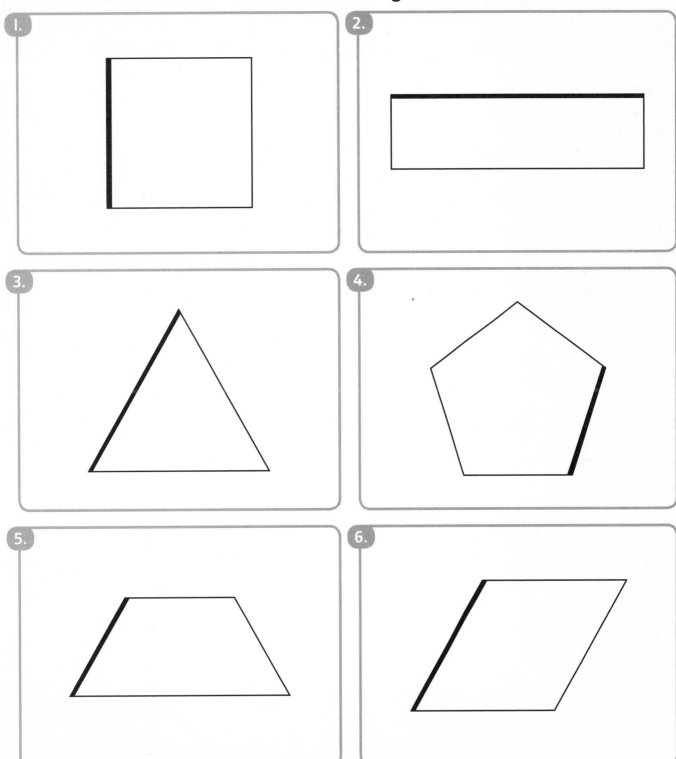

☐ Circle the polygons that have all sides the same length.

Some quadrilaterals have special names.

A **rhombus** has 4 sides of the same length.

rhombuses                                              not rhombuses

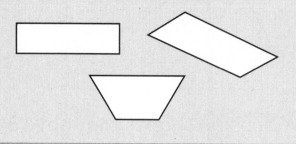

☐ Use a ruler to measure the sides.

☐ Is it a rhombus? If it is, draw ✓ inside the rhombus.

**7.**
_____ cm        _____ cm

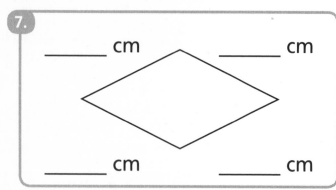

_____ cm        _____ cm

**8.**
_____ cm        _____ cm

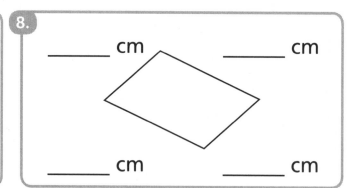

_____ cm        _____ cm

**9.**
_____ cm

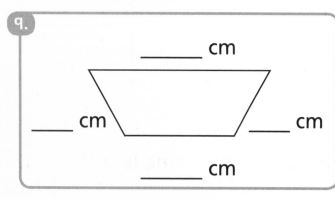

_____ cm        _____ cm

_____ cm

**10.**
_____ cm        _____ cm

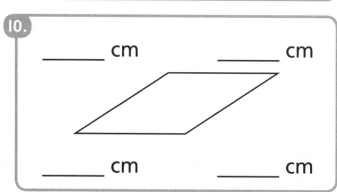

_____ cm        _____ cm

☐ This shape is not a rhombus. How do you know?

**11.**

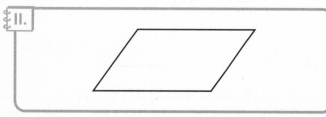

**12.**

**Geometry 2-6**

# G2-7 Polygons (Advanced)

☐ Color the shapes that have all sides the same length.

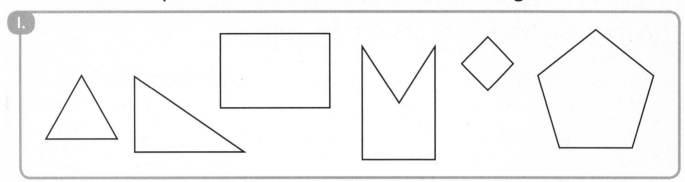

☐ Color the shapes that have 3 sides.

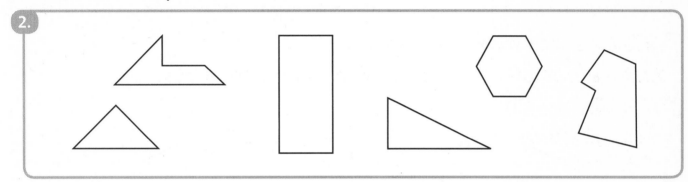

☐ Color the shapes that have 1 or more square corners.

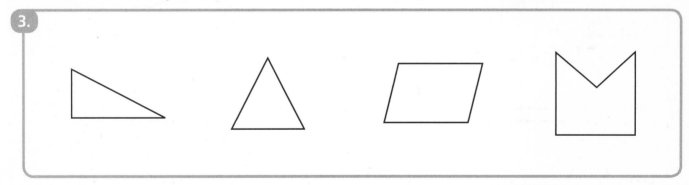

☐ **BONUS:** Color the shapes that have 4 sides the same length.

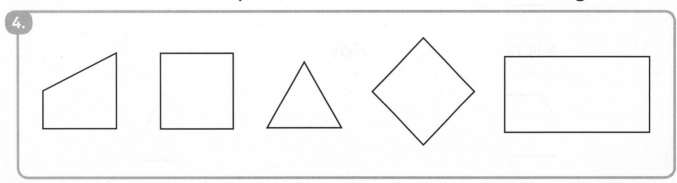

☐ Cross out the shapes that have **all** square corners.

☐ Color the shapes that have **no** square corners.

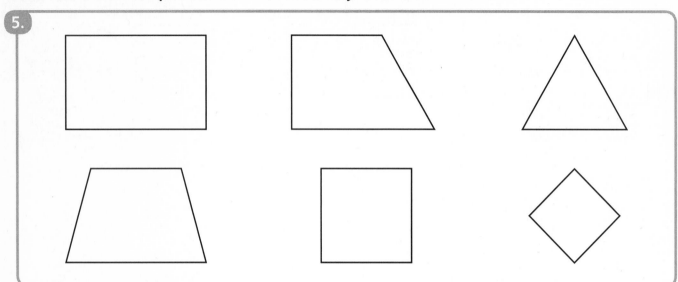

5.

☐ Measure the sides.

☐ Color the shape that has all sides the same length.

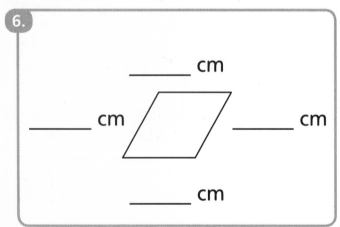

6.

_____ cm

_____ cm

_____ cm

_____ cm

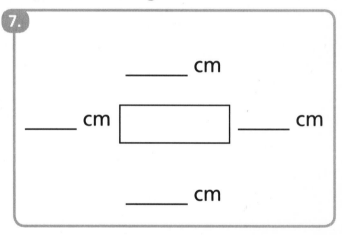

7.

_____ cm

_____ cm

_____ cm

_____ cm

8.
Draw a rhombus.
Draw a shape with 4 sides that is not a rhombus.
Explain why it is not a rhombus.

9.
Draw a rectangle that is not a square.
How do you know it is not a square?

# G2-8 Cubes

| cubes | not cubes |
|-------|-----------|
| 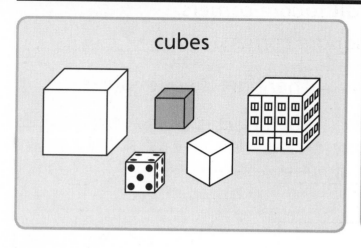 | 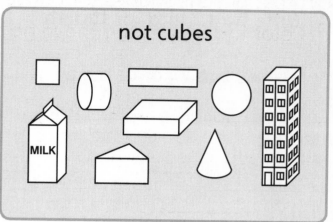 |

☐ Circle the cubes.

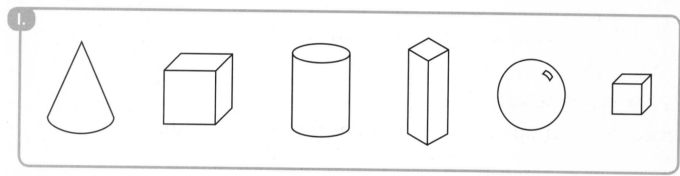

1.

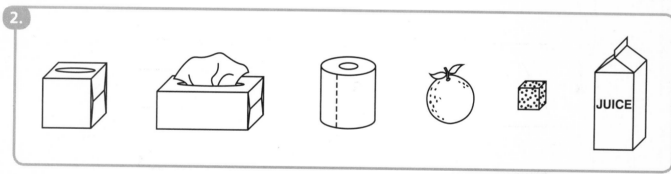

2.

3. Explain why a ▭ is not a cube.

Use the words **square** and **rectangle**.

4. **BONUS**

The shaded part is called a face.

Count the number of faces on a cube.

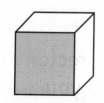

# G2-9 Fractions

☐ Write **half**, **third**, or **fourth**.

**1.**
There are 2 equal parts.

Each part is a
___half___.

**2.**
There are 3 equal parts.

Each part is a
_____.

**3.**
There are 4 equal parts.

Each part is a
_____.

**4.**
There are 2 equal parts.

Each part is a
_____.

**5.**
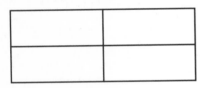

Each part is a _____.

**6.**

Each part is a _____.

**7.**
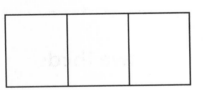

Each part is a _____.

**8.**

Each part is a _____.

**9.**
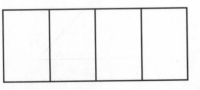

Each part is a _____.

**10.**
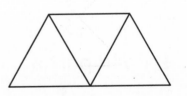

Each part is a _____.

## Color the fraction.

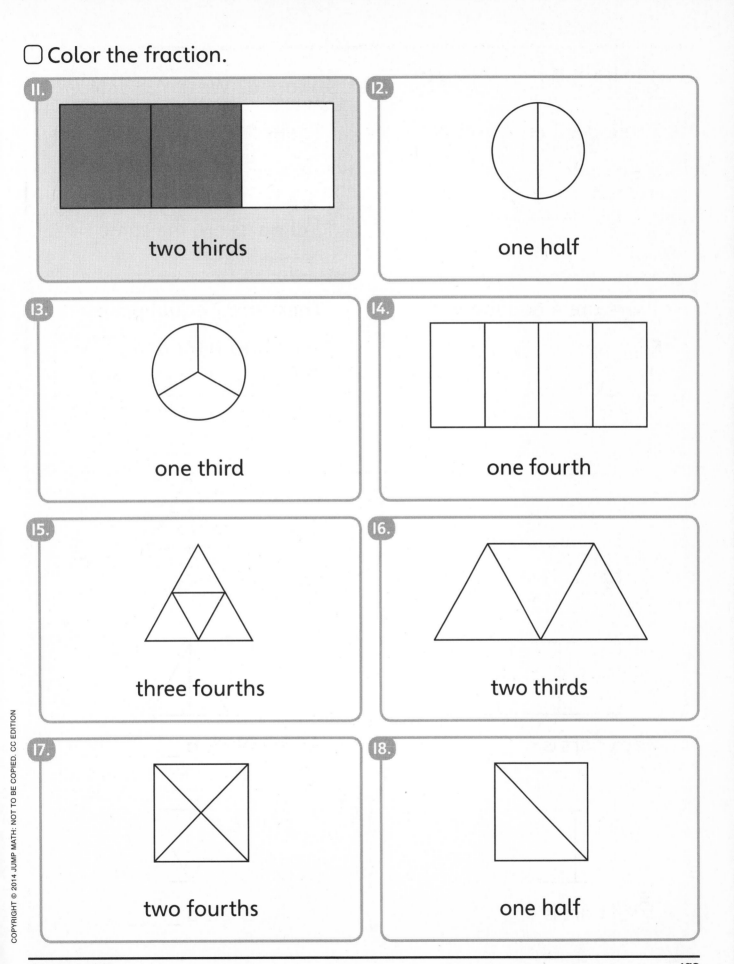

**11.** two thirds

**12.** one half

**13.** one third

**14.** one fourth

**15.** three fourths

**16.** two thirds

**17.** two fourths

**18.** one half

□ Does the picture have three fourths shaded? Write **Yes** or **No**.

**19.**

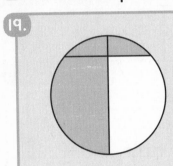

_No_

- ☑ 3 parts are shaded.
- ☑ There are 4 parts in total.
- ☒ All parts are the same size.

**20.**

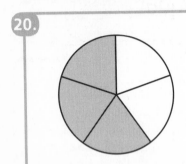

_____

- ☐ 3 parts are shaded.
- ☐ There are 4 parts in total.
- ☐ All parts are the same size.

**21.**

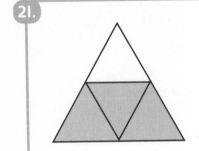

_____

- ☐ 3 parts are shaded.
- ☐ There are 4 parts in total.
- ☐ All parts are the same size.

**22.**

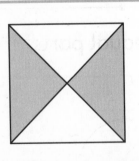

_____

- ☐ 3 parts are shaded.
- ☐ There are 4 parts in total.
- ☐ All parts are the same size.

**23.**

Does the picture have three fourths shaded?
Explain how you know.

# G2-10 More Fractions

☐ Connect the dots to divide the shape into equal parts.

☐ Count the equal parts.

**1.**

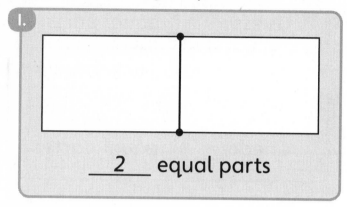

___2___ equal parts

**2.**

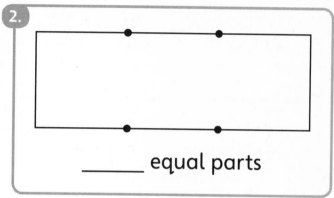

_____ equal parts

**3.**

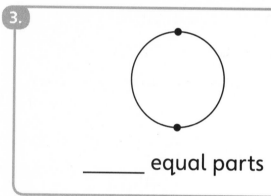

_____ equal parts

**4.**

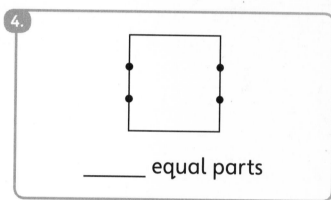

_____ equal parts

**5.**

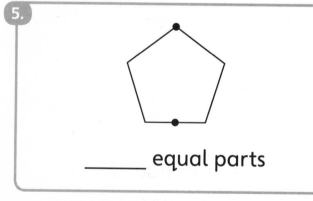

_____ equal parts

**6.**

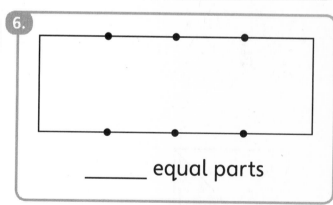

_____ equal parts

**7.**

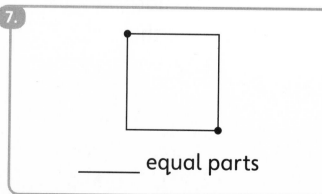

_____ equal parts

**8. BONUS**

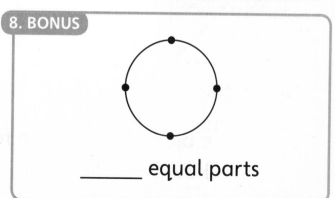

_____ equal parts

☐ Guess how many equal parts the shape will have.
☐ Connect the dots and count the equal parts to check.

**9.**
Guess ___3___ equal parts

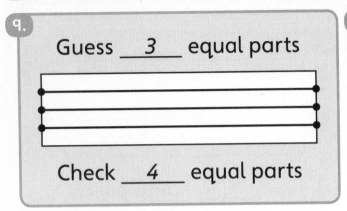

Check ___4___ equal parts

**10.**
Guess _____ equal parts

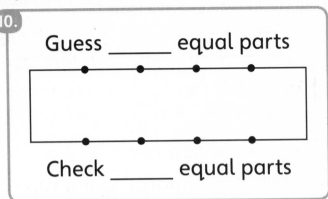

Check _____ equal parts

☐ Divide the shape in two different ways.
☐ How many equal parts are there?

**11.**

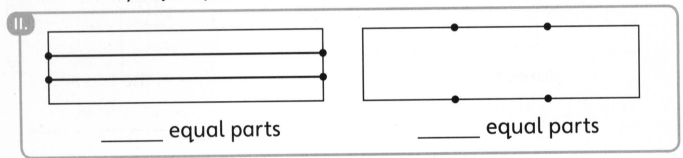

_____ equal parts          _____ equal parts

☐ Draw a line to divide the square into halves.
  Show three different ways.

**12.**

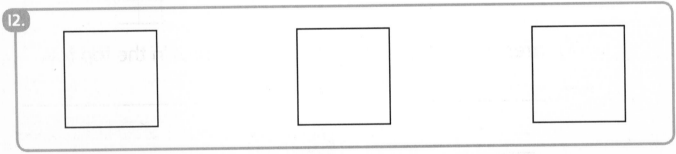

**13.**
Alexa thinks she got the same fraction
of a sandwich as Ben.

Is she correct? Explain.

Alexa's piece     Ben's piece

# G2-II Dividing a Rectangle

☐ Count the small squares.

**1.**

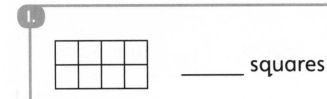 _____ squares

**2.**

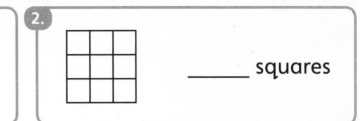 _____ squares

☐ Shade the top row.
☐ Count the squares in the top row.
☐ Write an addition for the total number of squares.

**3.**

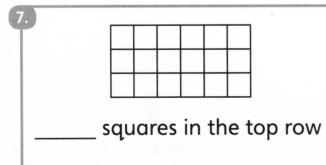 ← row

__5__ squares in the top row

__5 + 5 + 5_____

**4.**

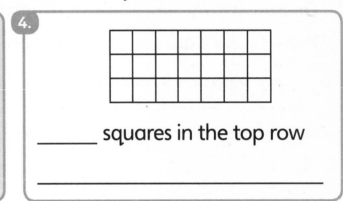

_____ squares in the top row

_____

**5.**

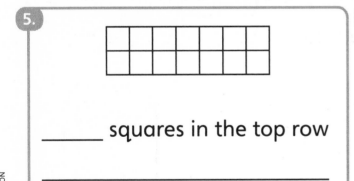

_____ squares in the top row

_____

**6.**

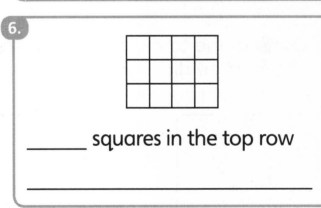

_____ squares in the top row

_____

**7.**

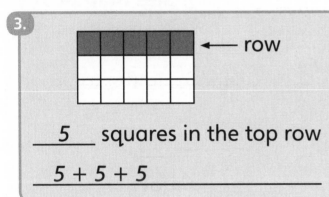

_____ squares in the top row

_____

**8.**

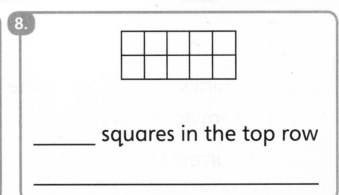

_____ squares in the top row

_____

☐ Count the squares in the top row.

☐ Skip count to find how many squares in total.

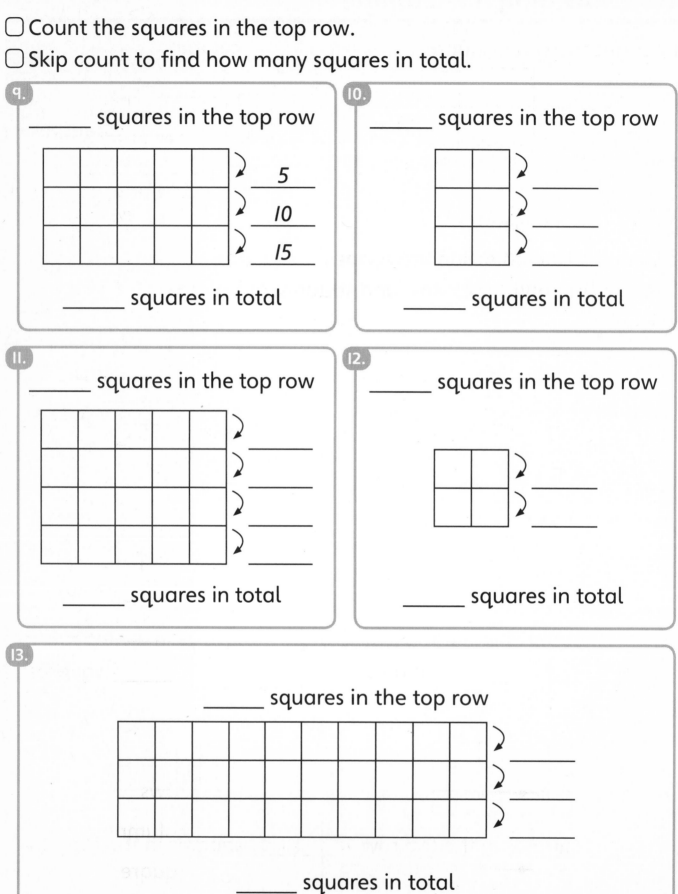

**9.**

_____ squares in the top row

5
10
15

_____ squares in total

**10.**

_____ squares in the top row

_____
_____
_____

_____ squares in total

**11.**

_____ squares in the top row

_____
_____
_____
_____

_____ squares in total

**12.**

_____ squares in the top row

_____
_____

_____ squares in total

**13.**

_____ squares in the top row

_____
_____
_____

_____ squares in total

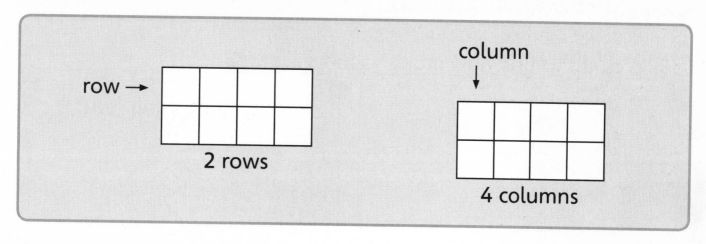

row → 2 rows

column
↓
4 columns

☐ Join the dots to divide the shape into squares.
☐ Count the rows, columns, and squares.

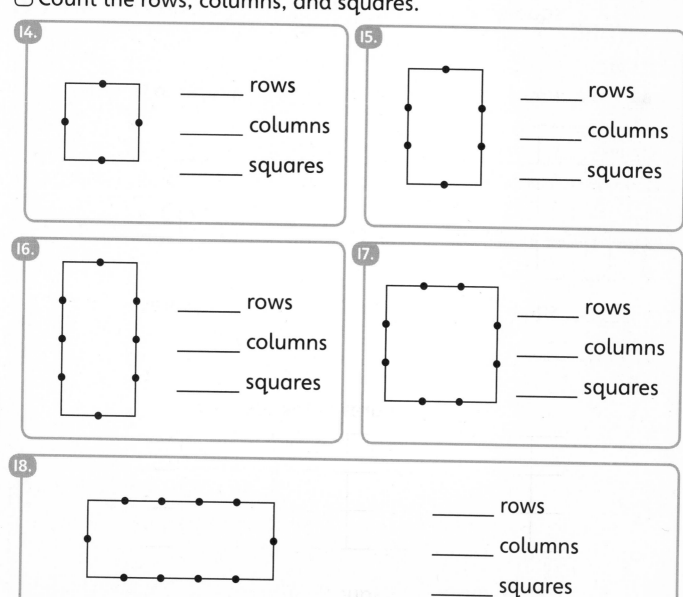

**14.**
_____ rows

_____ columns

_____ squares

**15.**
_____ rows

_____ columns

_____ squares

**16.**
_____ rows

_____ columns

_____ squares

**17.**
_____ rows

_____ columns

_____ squares

**18.**
_____ rows

_____ columns

_____ squares

# G2-I2 Problems and Puzzles

☐ Answer the questions.

**1.**

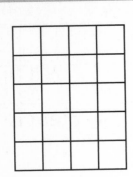

How many small squares in total? Write an addition sentence.

_____

**2.**

How many pentagons?

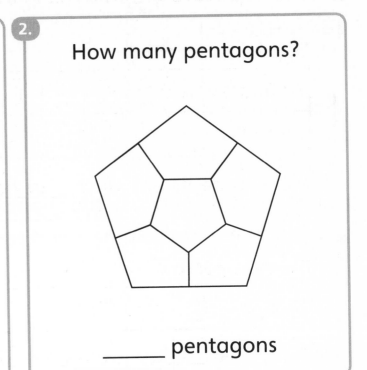

_____ pentagons

☐ Draw a shape with 4 sides that has no square corners.

**3.**

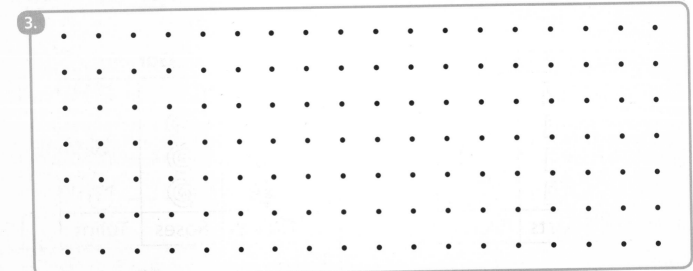

**4.**

Abdul and Beth share a sandwich. Each person gets an equal part. What fraction did Beth get?

Draw a picture to show your answer.

# MD2-44 Picture Graphs

☐ Use the **picture graph** to fill in the blanks.

**1.**

**Lunch**

| At home | ☺ | ☺ | ☺ | ☺ | ☺ | ☺ |
|---------|---|---|---|---|---|---|
| At school | ☺ | ☺ | ☺ | ☺ | | |

___6___ eat at home.

___4___ eat at school.

More students eat lunch ___*at home*___ .

**2.**

**Gloves or Mitts**

| Gloves | 🖐 | 🖐 | 🖐 | | |
|--------|---|---|---|---|---|
| Mitts | 🧤 | 🧤 | 🧤 | 🧤 | 🧤 |

_____ wear gloves.

_____ wear mitts.

More students wear _____ .

**3.**

### Sara's Clothes

| Shorts | Skirts |
|--------|--------|

Sara has _____ shorts.

Sara has _____ skirts.

Sara has fewer _____

than _____ .

**4.**

### Marco's Garden

| Daisies | Roses | Tulips |
|---------|-------|--------|

There are _____ daisies,

_____ roses, and _____ tulips.

Marco has the same number

of _____ and _____ .

Use the picture graph to fill in the blanks.

**5.**

### Shoes

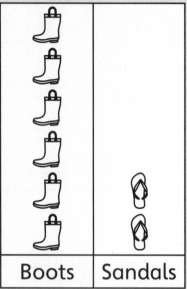

| Boots | Sandals |

| 6 | _boots_ |

| 2 | _sandals_ |

| 6 | − | 2 | = _4_ |

___4___ fewer people wear sandals than boots.

**6.**

### Students

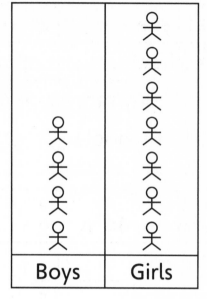

| Boys | Girls |

[ ] _____

[ ] _____

[ ] − [ ] = _____

There are _____ more girls than boys.

**7.**

**Birds We Saw**

| Pigeons | 🐦 🐦 🐦 |  |  |  |
| Robins | 🐦 🐦 🐦 🐦 🐦 |  |  |

[ ]  _____
[ ]  _____

[ ] − [ ] = _____

We saw _____ fewer pigeons than robins.

Use the graph to answer the questions.

**8.**

**Lunch in Ms. Alba's Class**

| At home | H | H | H | H | H | H | H | H | H | H | H | H |
|---|---|---|---|---|---|---|---|---|---|---|---|---|
| At school | S | S | S | S | S | S | S | | | | | |

How many more students eat lunch at home

than at school? _____

**9.**

**Lunch in Mr. Wong's Class**

| At home | | | | | | | | | | |
|---|---|---|---|---|---|---|---|---|---|---|
| At school | S | S | S | S | S | S | S | S | S | S |

There are 17 students in Mr. Wong's class.

How many students eat lunch at school? _____

How many eat lunch at home? _____

Complete the graph. Write **H** to show the data.

Use data from the graphs above to complete the graph.

**10.**

**Lunch at Home**

| Ms. Alba's class | | | | | | | | | | | |
|---|---|---|---|---|---|---|---|---|---|---|---|
| Mr. Wong's class | | | | | | | | | | | |

Which class has more students eating at home?

_____

How many more? _____

☐ Draw 😊 to show the data.

☐ Answer the questions.

**11.**

**Favorite Ball Game**

| Baseball | 😊 | 😊 | 😊 | 😊 | 😊 | 5 like baseball. |
| Basketball | | | | | | 3 like basketball. |
| Soccer | | | | | | 4 like soccer. |

How many more students like baseball than soccer? _____

**12.**

**Shoes We Wear**

| Dress shoes | | | | | | 2 wear dress shoes. |
| Boots | | | | | | 5 wear boots. |
| Sandals | | | | | | 4 wear sandals. |

How many fewer students wear sandals than boots? _____

**13.**

Clara asked her friends where they will be during the holiday.

**Holiday Plans**

| Camp | | | | | | 5 will go camping. |
| Cottage | | | | | | 2 will go to a cottage. |
| Home | | | | | | 3 will stay at home. |

How many friends did Clara ask? _____

How many friends will not stay at home? _____

# MD2-45 Drawing Tallies and Picture Graphs

☐ Write the number or draw the **tally**.

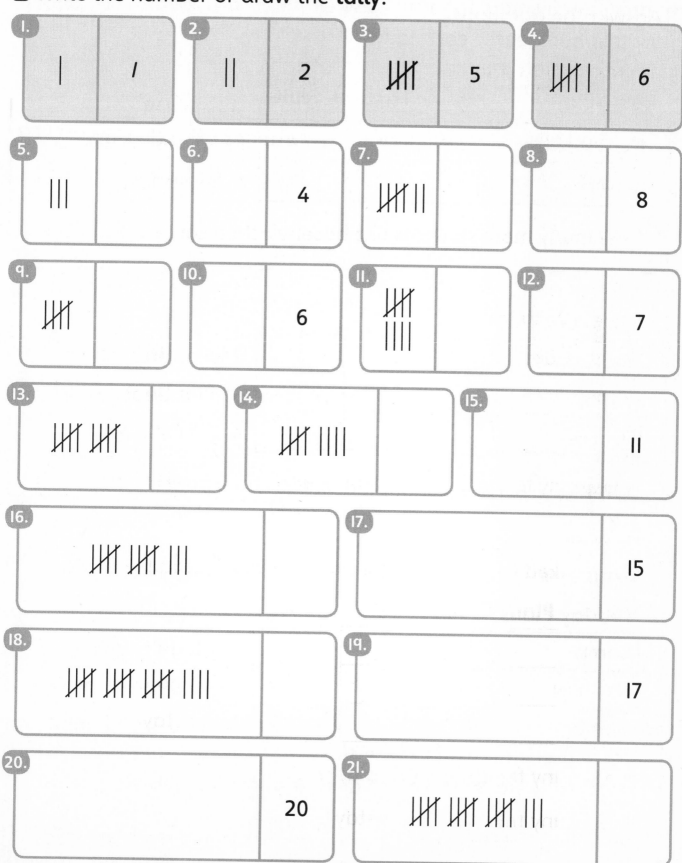

1. | | 1

2. || 2

3. ||||  5

4. |||| | 6

5. ||| 

6. 4

7. |||| || 

8. 8

9. |||| 

10. 6

11. |||| |||| 

12. 7

13. |||| |||| 

14. |||| |||| 

15. 11

16. |||| |||| ||| 

17. 15

18. |||| |||| |||| |||| 

19. 17

20. 20

21. |||| |||| |||| |||

Sam asked students in his class to name their favorite fruit.

He drew a tally mark for each answer.

☐ Write a number for each tally.

**22.**

**Favorite Fruit**

| Apples | Bananas | Oranges | Pears |
|--------|---------|---------|-------|
| 卌 ‖ | 卌 卌 | 卌 ∣ | 卌 |
| 7 | | | |

☐ Use the tallies above to answer the questions.

**23.**

How many students liked oranges best? _____

How many students liked pears the best? _____

Which was the most popular fruit? _____

Which was the least popular fruit? _____

How many students answered Sam's question? _____

How many more students liked bananas than pears? _____

**24. BONUS**

There are 30 students in the class.

How many students did not answer Sam's question?

_____

☐ Make a picture graph using the data.

☐ What **symbol** did you use to show the data?

**25.**

### Bird Watching

| Student | Birds Seen |
|---------|------------|
| Anwar | \|\|\|\|\| |
| Carlos | \|\|\| |
| Emma | \|\| |
| Grace | \|\|\|\| \| |

### Bird Watching

| Student | Birds Seen | | | | | |
|---------|---|---|---|---|---|---|
| Anwar | ○ | ○ | ○ | ○ | ○ | |
| Carlos | | | | | | |
| Emma | | | | | | |
| Grace | | | | | | |

Each _____ stands for one bird.

How many more birds did Grace see than Emma? _____

**26.**

Cathy asked her classmates which animal is their favorite.

5 chose elephants.                    4 chose giraffes.

3 chose lions.                         6 chose tigers.

### Favorite Animal

| Animal | Number of Students | | | | | |
|--------|---|---|---|---|---|---|
| Elephants | | | | | | |
| Giraffes | | | | | | |
| Lions | | | | | | |
| Tigers | | | | | | |

Each _____ stands for one animal.

# MD2-46 Bar Graphs

◯ How many?

**1.**

### Pets We Own

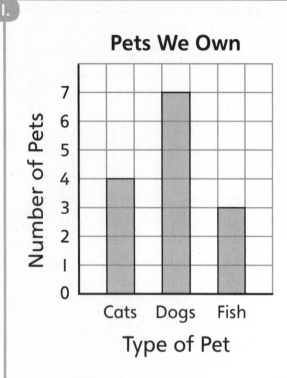

The bar for cats is __4__ long.

We own __4__ cats.

The bar for dogs is _____ ▢ long.

We own _____ dogs.

The bar for fish is _____ ▢ long.

We own _____ fish.

**2.**

### Flowers in Ethan's Garden

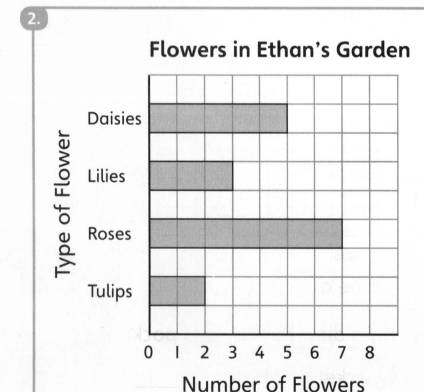

Ethan has:

_____ daisies

_____ lilies

_____ roses

_____ tulips

Use the bar graph to answer the questions.

**3.**

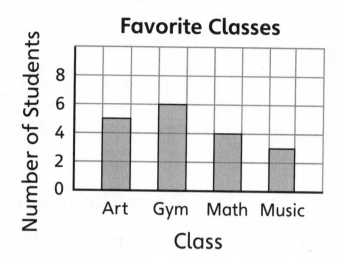

Which class is most popular? _____

Which class is least popular? _____

**4.**

**Coins in Ed's Pocket**

How many more quarters than dimes are in Ed's pocket? _____

How many coins are in Ed's pocket altogether? _____

# MD2-47 Making Bar Graphs

☐ Use the data to complete the bar graph.

☐ Answer the questions.

1.

**Our Shoes**

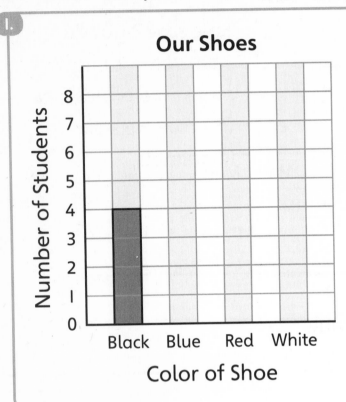

| Color | Number |
|-------|--------|
| Black | 4 |
| Blue | 7 |
| Red | 8 |
| White | 3 |

Which color of shoe is most

common? _____

Which color of shoe is least

common? _____

2.

**Kate's Snacks**

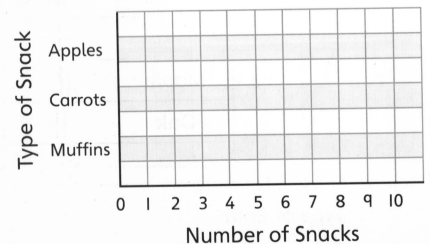

Kate has 5 apples, 7 carrots, and 3 muffins.

How many more carrots than apples does Kate have? _____

Complete the graph.

**3.**

### Zoo Animals that Jake Saw

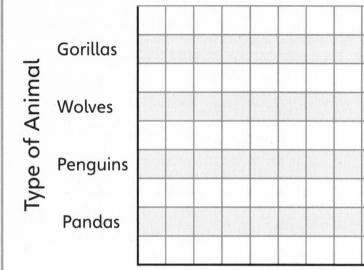

Type of Animal

Gorillas

Wolves

Penguins

Pandas

0  I  2  3  4  5  6  7  8

Number of Animals

Jake saw 3 gorillas, 5 wolves, 4 penguins, and some pandas at the zoo. He saw 14 animals.

How many pandas did he see? _____

**4.**

### Trees in Schoolyard

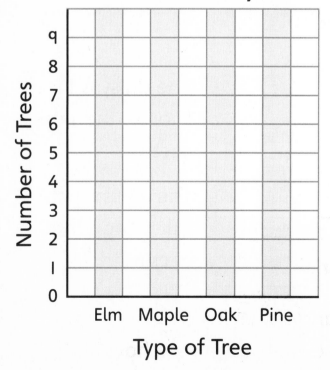

Number of Trees

q
8
7
6
5
4
3
2
I
0

Elm  Maple  Oak  Pine

Type of Tree

| Type of Tree | Number of Trees |
|---|---|
| Elm | ||| |
| Maple | ||||| | |
| Oak | || |
| Pine | ||||| ||| |

How many trees are not pines? _____

☐ Complete the graph.

**5.**

**Our Pets**

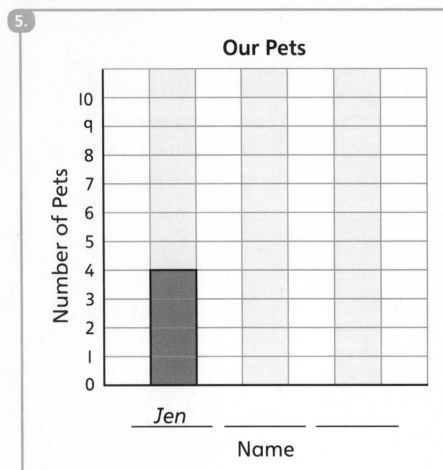

Don has 2 more pets than Jen.

Kyle has 3 fewer pets than Don.

How many pets do Don, Jen, and Kyle have altogether?

_____

**6. BONUS**

**Heights of Plants**

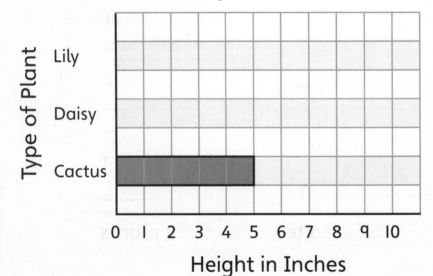

The daisy is 1 inch shorter than the cactus.

The lily is 1 inch taller than the cactus.

**Measurement and Data 2-47**

# MD2-48 Comparing Graphs (Advanced)

| Plant | Height |
|-------|--------|
| Bean | 3 inches |
| Kale | 5 inches |
| Beet | 2 inches |
| Pea | 5 inches |

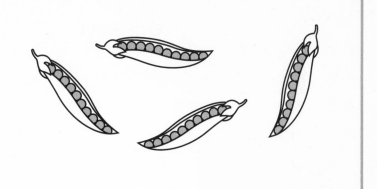

☐ Show the data two ways.

**1.**

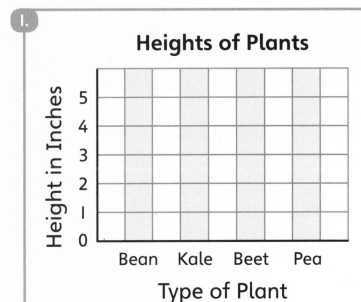

**Heights of Plants**

Height in Inches

5
4
3
2
1
0

Bean  Kale  Beet  Pea

Type of Plant

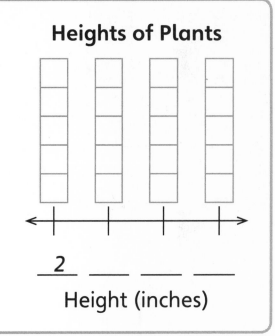

**Heights of Plants**

$\underset{2}{\underline{\quad}} \quad \underline{\quad} \quad \underline{\quad} \quad \underline{\quad}$

Height (inches)

☐ Use the graphs to answer the questions.

**2.**

Which graph is easier to make? _____

Can you see the type of plant in the line plot? _____

Which graph makes it easier to tell how many plants

are 5 inches tall? _____